CITES Orchid Checklist

Volume 1

For the genera:

Cattleya, Cypripedium, Laelia, Paphiopedilum, Phalaenopsis, Phragmipedium, Pleione and *Sophronitis*

including accounts of:

Constantia, Paraphalaenopsis and *Sophronitella*

CITES Orchid Checklist

Volume 1

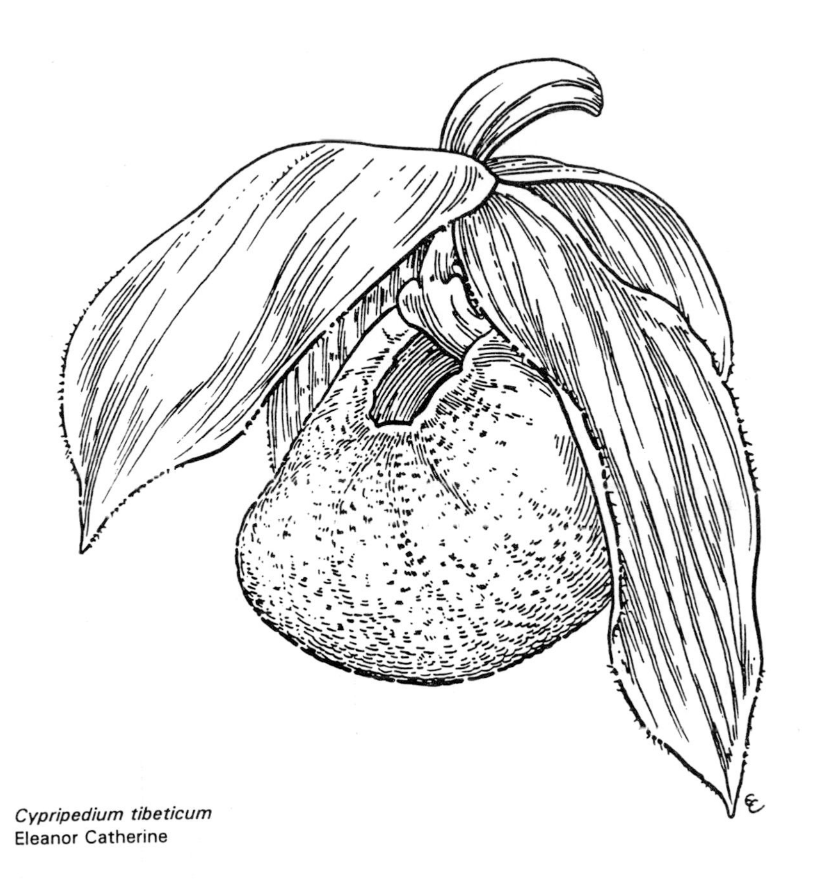

Cypripedium tibeticum
Eleanor Catherine

Compiled by:

Jacqueline A Roberts, Clive R Beale, Johanna C Benseler, H Noel McGough & Daniela C Zappi

Assisted by a selected panel of orchid experts

Royal Botanic Gardens, Kew

First published in 1995

General editor of series J.M. Lock

ISBN 0 947643 87 7

Compiled with the financial assistance of
the CITES Nomenclature Committee
and the Royal Botanic Gardens, Kew

Cover design by Media Resources RBG Kew
Printed and bound by Whitstable Litho Ltd, Whitstable, Kent

FOREWORD

The orchid family, being one of the largest in the plant kingdom, is certainly the largest single taxon included in the CITES appendices. In number of species it constitutes more than half of the total currently covered by CITES. Because of its size it is unfortunately the source of much confusion regarding the names used in trade. This makes it very difficut to adequately analyse the annual reports of the Parties, in order to establish whether the quantities traded are detrimental to the survival of the species concerned. In addition, because of the lack of adequate reference works, one frequently encounters misspelled names, making it difficult to determine which species is actually referred to. Moreover, by using incorrectly spelled names or even fake names, traders try to circumvent existing trade controls or trade bans.

The CITES Secretariat was therefore very pleased by the fact that the proposal of the Plants Committee to develop a checklist for the most commonly traded orchid taxa was approved at the eighth meeting of the Conference of the Parties, and that adequate funding had been allocated in the Secretariat's budget. This volume will be the first of many. Parties should use this checklist when issuing permits and preparing their annual reports. If this is done, the strongly needed, and correct, information will be available for the significant trade studies that the Plants Committee is currently initiating.

When the Cactaceae Checklist was published in 1992, I wrote that "for many years the fate of endangered plant species in trade received little attention from the Parties to CITES". Now we witness a change for the better, owing to the active work of the Plants Committee since that time, and the tireless efforts of the CITES Secretariat. The CITES Orchid Checklist is a next step in the further improvement of CITES implementation for plants. The Secretariat expects that those involved in CITES implementation will make good use of this checklist.

IZGREV TOPKOV
Secretary General, CITES
April 1995

Acknowledgements

The compilers would like to thank colleagues at the Royal Botanic Gardens, Kew and the following orchid experts for their help with the preparation of the checklist for publication. All suggestions for amendments were gratefully noted and included at the discretion of the compilers. We would particularly like to thank Dr Alec Pridgeon and Dr Phillip Cribb for running a critical eye over the final version of the text.

International Panel of Orchid Experts

Dr J Atwood	USA
Mr I Butterfield	UK
Dr S-C Chen	China
Dr E Christenson	USA
Mr J Comber	UK
Dr P Cribb	UK
Mr W Frosch	Germany
Mr O Gruß	Germany
Dr K W Tan	Singapore
Prof H Koopowitz	USA
Mr H Perner	Germany
Dr A Pridgeon	UK
Dr G Romero	USA
Dr S Schneckenburger	Germany
Dr K Senghas	Germany
Dr H Valmayor	Philippines
Dr C Withner	USA
Mr J Wood	UK

CONTENTS

Preamble

1. Background 1
2. Computer aspects 2
3. Compilation procedures 2
4. Conservation 3
5. How to use the checklist 3
6. Conventions employed in parts I, II and III 3
7. Number of names entered for each genus 4
8. Geographical areas 4
9. Orchidaceae controlled by CITES 4
10. Bibliography 4

Part I: ORCHIDACEAE BINOMIALS IN CURRENT USAGE: Ordered Alphabetically on All Names (includes distribution) 7

Part II: ORCHIDACEAE BINOMIALS IN CURRENT USAGE: Ordered Alphabetically on Accepted Names 35

- *Cattleya* 37
- *Constantia* 45
- *Cypripedium* 47
- *Laelia* 53
- *Paphiopedilum* 61
- *Paraphalaenopsis* 73
- *Phalaenopsis* 75
- *Phragmipedium* 87
- *Pleione* 91
- *Sophronitella* 95
- *Sophronitis* 97

Part III: COUNTRY CHECKLIST 99

Annex I: IUCN Red List Categories 117

CITES CHECKLIST - ORCHIDACEAE

PREAMBLE

1. Background

The 1992 Conference of the Parties to the Convention on International Trade in Endangered Species of Wild Fauna and Flora (CITES) adopted Resolution Conf. 8.19 which called for the production of a standard reference to the names of Orchidaceae.

The Vice-Chairman of the CITES Nomenclature Committee was charged with the responsibility of co-ordinating the input needed to produce such a reference.

The orchid genera identified as priorities in the *Review of Significant Trade in Species of Plants included in Appendix II of CITES* (CITES Doc. 8.31) would be treated first. The checklists (or parts thereof) as they came available would be put to the Conference of the Parties for approval.

At its third meeting (Chiang Mai, Thailand, November 1992) the Plants Committee extensively discussed a proposal by the Vice-Chairman of the Nomenclature Committee regarding the possible mechanisms to develop the Standard Reference. The Plants Committee endorsed a procedure by which compilations from the available literature, made on a central database, would be circulated to a panel of international experts for consultation and final decisions on the valid names to be used for the taxa concerned.

Based on the recommendations of the Plants Committee and on those of Resolution Conf. 8.19, the CITES Secretariat established a Memorandum of Understanding with the Royal Botanic Gardens, Kew for the preparation of this reference. The work started on the 1 July 1993. Recommendation 6 of the *Review of Significant Trade of Species of Plants included in Appendix II of* CITES outlined the following genera as priorities:

Aerangis, Angraecum, Ascocentrum, Bletilla, Brassavola, Calanthe, Catasetum Cattleya, Coelogyne, Comparettia, Cymbidium, Cypripedium, Disa, Dracula, Encyclia, Epidendrum, Laelia, Lycaste, Masdevallia, Miltonia, Miltoniopsis, Odontoglossum, Oncidium, Paphiopedilum, Paraphalaenopsis, Phalaenopsis, Phragmipedium, Renanthera, Rhynchostylis, Rossioglossum, Sophronitis, Vanda and *Vandopsis*.

The Memorandum of Understanding stated that completed and verified checklists of the following groups should be prepared for consideration by the ninth meeting of the Conference of the Parties:

Cattleya, Cypripedium, Laelia, Paphiopedilum, Phalaenopsis, Phragmipedium, Pleione and *Sophronitis*.

Based on the agreed formula, work commenced at Kew. In the course of this work the panel also supplied information on *Constantia*, *Paraphalaenopsis* and *Sophronitella*, and for this reason they are also included in this publication. The Vice-Chairman of the Nomenclature Committee reported on progress at the fourth (Brussels, Belgium, September 1993) and fifth (San Miguel de Allende, Mexico, May 1994) meetings of the Plants Committee. The 9th Conference of the Parties approved the continuing process and adopted the initial checklist and its updates, accepted by the Nomenclature Committee, as a guideline when making reference to the species names of the genera concerned. The publication of this, the first part of the standard reference fulfils the commitment undertaken by the Royal Botanic Gardens, Kew.

Preamble

2. Computer aspects

Hardware: The database was set up on a Compaq LTE Lite 4/25C laptop computer using ALICE software.

Database system: The ALICE database system was used to handle the data collection. "ALICE handles distribution, uses, common names, descriptions, habitats, synonymy, bibliography and other classes of data for species, subspecies or varieties. It allows users to design their own reports for checklists, studies, monographs or conservation lists for example.

The ALICE System is a family of programs which operate on ALICE databases. Each has a different purpose.

ALICE: The main program for the database developer: database design, data capture and editing. Reports use simple pre-defined formats.

ATEXT: To incorporate, edit and view free-text species descriptions.

AQUERY: Flexible data retrieval with a simple interface suitable even for those with only superficial knowledge of the database.

AWRITE: For designing and printing reports.

NVIEW: For nomenclatural exploration of a database.

ALEX: To export data subsets into other program formats including: SDF, dBASE, Tabular or Fixed-length fields, DELTA and XDF.

ASLICE: To export data subsets as new independent ALICE databases.

AMIE: For routine maintenance of ALICE databases.

ALICE is currently written in XBASE and C. ALICE can run under DOS, Windows, VMS, UNIX and XENIX. ALICE can also run on Oracle, Recital and Ingres." (ALICE Software Partnership, November 1991).

3. Compilation procedures

- Primary references were identified by orchid specialists based at the Royal Botanic Gardens, Kew.
- A Panel of Orchid Experts was established in order to review each stage of the checklist.
- Information was entered into the ALICE taxonomic database and a preliminary report produced.
- Preliminary reports for each genus were distributed to the Panel of Orchid Experts for their comments on any additions or amendments needed.
- Additions and amendments returned from the Panel members were entered into the database. These were linked to a reference contained in the bibliography at the end of the report on each genus. (Not included in this checklist, but copies held for reference at RBG, Kew).
- This sequence was repeated three times for each genus to allow full consultation with the Panel.
- A preliminary checklist was compiled for review at the 9th CITES Conference of the Parties (COP9).
- Additions and amendments subsequent to COP9 were added to the database.

- Format for publication was agreed with the CITES Secretariat and reports generated using AWRITE and prepared for camera-ready copy using Microsoft Word for Windows version 6.

4. Conservation

During the consultation process with the Panel of Orchid Experts, information was also requested on the conservation status of the species concerned. Copies of the present and proposed new IUCN categories of threat were distributed to the Panel for their use. Unfortunately, there were limited returns from the Panel and the information has not been included in this checklist. Annex I threfore provides a copy of the recently approved IUCN Red List Categories. We hope that this will help make the new categories available to orchid experts to use in the field.

5. How to use the checklist

It is intended that this Checklist be used as a quick reference for checking accepted names, synonymy and distribution. The reference is therefore divided into three main parts:

Part I: ORCHIDACEAE BINOMIALS IN CURRENT USAGE - ALL NAMES

An alphabetical list of all accepted names and synonyms included in this checklist - a total of 1349 names.

Part II: ORCHIDACEAE BINOMIALS IN CURRENT USAGE - ACCEPTED NAMES

Separate lists for each genus. Each list is ordered alphabetically by the accepted name and details are given on current synonyms and distribution.

Part III: COUNTRY CHECKLIST

Accepted names from all genera included in this checklist are ordered alphabetically under country of distribution.

6. Conventions employed in parts I, II and III

a) Accepted names are presented in **bold roman** type.
Synonyms are presented in *italic* type.

b) Where a synonym occurs twice, but refers to two different accepted names, eg, *Cattleya pumila,* for both **Cattleya lawrenceana** and **Laelia pumila**, the name with an asterix is the species most likely to be encountered in trade. For example:

All Names	Accepted Name
Cattleya pterocarpa	**Sophronitis pterocarpa**
Cattleya pumila	**Cattleya lawrenceana**
Cattleya pumila	**Laelia pumila***
Cattleya pumila var. *major*	**Laelia spectabilis**
Cattleya purpurata	**Laelia purpurata**

*Species most likely to be in trade (in this example, **Laelia pumila**).

c) Where an accepted name and a synonym are the same, but refer to different species, eg, **Cypripedium wardii** and *Cypripedium wardii* (**Paphiopedilum wardii**), the name with an asterix is the species most likely to be seen in trade. For example:

All Names	Accepted Name
Cypripedium waltersianum	**Paphiopedilum appletonianum**
Cypripedium wardii	
Cypripedium wardii	**Paphiopedilum wardii***
Cypripedium wilsonii	**Cypripedium fasciolatum**
Cypripedium wolterianum	**Paphiopedilum appletonianum**

*Species most likely to be in trade (in this example, **Paphiopedilum wardii**).

NB: In examples b) and c) it is necessary to double-check by reference to the distribution as detailed in Part II. For instance, in the example b), if it was known that the plant in question came from Myanmar it is likely to be **Paphiopedilum wardii** as **Cypripedium wardii** is found in China.

d) Natural hybrids have been included in the checklist and are indicated by the multiplication sign ×. They are arranged alphabetically within the lists.

7. Number of names entered for each genus:
Cattleya (Accepted: 50, Synonyms: 169); *Constantia* (Accepted: 5, Synonyms: 2); *Cypripedium* (Accepted: 49, Synonyms 77); *Laelia* (Accepted: 59, Synonyms: 126); *Paphiopedilum* (Accepted: 82, Synonyms: 239); *Paraphalaenopsis* (Accepted: 5, Synonyms: 7); *Phalaenopsis* (Accepted: 81, Synonyms: 194); *Phragmipedium* (Accepted: 19, Synonyms: 68); *Pleione* (Accepted: 20, Synonyms: 62); *Sophronitella* (Accepted: 1, Synonyms: 2); *Sophronitis* (Accepted: 9, Synonyms: 23).

8. Geographical areas
Country names follow the United Nations standard as laid down in Country Names. *Terminology Bulletin* 1993. United Nations 345:1-151.

9. Orchidaceae controlled by CITES
The family Orchidaceae is listed on Appendix II of CITES. In addition the following taxa are listed on Appendix I at time of publication:

Cattleya trianaei
Dendrobium cruentum
Laelia jongheana
Laelia lobata
Paphiopedilum spp.
Peristeria elata
Phragmipedium spp.
Renanthera imschootiana
Vanda coerulea

10. Bibliography
Primary reference sources used in the compilation of checklists:

Banerji, M.L. & Pradhan, P. (1984). *The Orchids of Nepal Himalaya.* J. Cramer, in der A.R. Gantner Verlag Kommanditgesellschaft.

Braem, G.J. (1984). *Cattleya. The Brazilian bifoliate Cattleyas.* Brücke-Verlag Schmersow, Hildesheim.

Braem, G.J. (1986). *Cattleya vol II. The unifoliate Cattleyas.* Brücke-Verlag Schmersow, Hildesheim.

Braem, G.J. (1988). *Paphiopedilum: a monograph of all tropical and subtropical Asiatic slipper-orchids.* Brücke-Verlag Schmersow, Hildesheim.

Chen, S.C. & Xi, Y.Z. (1987). Chinese Cypripediums in *Proceedings of the 12th World Orchid Conference, March 1987, Tokyo, Japan.* 12th World Orchid Conference, Inc., Japan.

Comber, J.B. (1990). *Orchids of Java.* The Bentham-Moxon Trust, Royal Botanic Gardens, Kew, UK.

Cribb, P.J. (1987). *The Genus Paphiopedilum.* A Kew Magazine Monograph. Royal Botanic Gardens, Kew in association with Collingridge, UK.

Cribb, P.J. (1994). *The Genus Cypripedium in Mexico and Central America.* Orquidea (Mex.) 13(1&2): 205-214.

Cribb, P.J. & Butterfield, I (1988). *The Genus Pleione.* Royal Botanic Gardens, Kew in association with Christopher Helm, UK.

Garay, L.A. (1979). *The Genus Phragmipedium.* Orchid Digest 43(4): 133-148.

Luer, C.A. (1975). *The Native Orchids of the United States and Canada, excluding Florida.* New York Botanical Garden, USA.

McCook, L.M. (1989). *Systematics of Phragmipedium.* Dissertation presented to Cornell University, USA.

Pabst, G.F.J. & Dungs, F. (1975). *Orchidaceae Brasilienses.* Vol. 1. Hagemann-Druck, Hildesheim.

Pabst, G.F.J. & Dungs, F. (1977). *Orchidaceae Brasilienses* Vol. 2. Hagemann-Druck, Hildesheim.

Pradhan, U.C. (1976). *Indian Orchids: guide to identification and culture.* Vol. 1. U.C. Pradhan, Kalimpong.

Pradhan, U.C. (1979). *Indian Orchids: guide to identification and culture.* Vol. 2. U.C. Pradhan, Kalimpong.

Seidenfaden, G & Smitinand, T. (1959 - 1964). *The Orchids of Thailand. A Preliminary List. Parts 1 - 4.* The Siam Society.

Seidenfaden, G. (1992). *The Orchids of Indochina* in Opera Bot. 114: 1-502.

Seidenfaden, G & Wood, J.J. (1992). *The Orchids of Peninsula Malaysia & Singapore.* Olsen & Olsen, Fredensborg.

Sweet, H.R. & Cribb, P.J. (1980). The Genus Phalaenopsis. *Orchid Digest: Orchids of the World.* Vol. 1.

Valmayor, H.L. (1984). *Orchidiana Philippiniana.* Eugenio Lopez Foundation, Manila.

Withner, C.L. (1988). *The Cattleyas and their Relatives* Vol. I. Portland, Or.: Timber Press, USA.

Withner, C.L. (1990). *The Cattleyas and their Relatives* Vol. II. Portland, Or.: Timber Press, USA.

Withner, C.L. (1993). *The Cattleyas and their Relatives* Vol III. Portland, Or.: Timber Press, USA.

Wood, J.J. & Cribb, P.J. (1994). *A Checklist of the Orchids of Borneo.* Royal Botanic Gardens, Kew, UK.

Zappi, D.C. (1995). Notes on the nomenclature of *Cattleya* and *Laelia* species (Orchidaceae) recognised by Withner (1988, 1990). *Kew Bulletin* (in press).

PART I: ORCHIDACEAE BINOMIALS IN CURRENT USAGE:
Ordered Alphabetically on All Names for the Genera:

Cattleya, Constantia, Cypripedium, Laelia, Paphiopedilum, Paraphalaenopis, Phalaenopsis, Phragmipedium, Pleione, Sophronitella and *Sophronitis*

ALPHABETICAL LISTING OF ALL NAMES FOR THE GENERA: *Cattleya, Constantia, Cypripedium, Laelia, Paphiopedilum, Paraphalaenopis, Phalaenopsis, Phragmipedium, Pleione, Sophronitella* and *Sophronitis*

ALL NAMES	ACCEPTED NAME
Amalias anceps	**Laelia anceps**
Amalias cinnabarina	**Laelia cinnabarina**
Angraecum album-majus	**Phalaenopsis amabilis**
Arietinum americanum	**Cypripedium arietinum**
Bletia acuminata	**Laelia rubescens**
Bletia albida	**Laelia albida**
Bletia anceps	**Laelia anceps**
Bletia autumnalis	**Laelia autumnalis**
Bletia boothiana	**Laelia lobata**
Bletia caulescens	**Laelia caulescens**
Bletia caulescens var. *libonis*	**Laelia caulescens**
Bletia cinnabarina	**Laelia cinnabarina**
Bletia cinnabarina var. *sellowii*	**Laelia cinnabarina**
Bletia crispa	**Laelia crispa**
Bletia crispilabia	**Laelia crispilabia**
Bletia elegans	**Laelia elegans**
Bletia flabellata	**Laelia xanthina**
Bletia flava	**Laelia flava**
Bletia furfuracea	**Laelia furfuracea**
Bletia grandiflora	**Laelia speciosa**
Bletia grandis	**Laelia grandis**
Bletia harpophylla	**Laelia harpophylla**
Bletia jongheana	**Laelia jongheana**
Bletia lobata	**Laelia lobata**
Bletia longipes	**Laelia longipes**
Bletia lucasiana	**Laelia lucasiana**
Bletia lundii	**Laelia lundii**
Bletia peduncularis	**Laelia rubescens**
Bletia perrinii	**Laelia perrinii**
Bletia praestans	**Laelia spectabilis**
Bletia pumila	**Laelia pumila**
Bletia purpurata	**Laelia purpurata**
Bletia rubescens	**Laelia rubescens**
Bletia rupestris	**Laelia crispata**
Bletia speciosa	**Laelia speciosa**
Bletia violacea	**Laelia rubescens**
Bletia xanthina	**Laelia xanthina**
Bletilla scopulorum	**Pleione scopulorum**
Broughtonia aurea	**Cattleya aurantiaca**
Calceolus candidus	**Cypripedium candidum**
Calceolus reginae	**Cypripedium reginae**
Catlaelia elegans	**Laelia elegans**
Cattleya aclandiae	
Cattleya aclandiae var. *schilleriana*	**Cattleya schilleriana**
Cattleya acuminata	**Laelia rubescens**
Cattleya albida	**Laelia albida**
Cattleya alexandrae	**Cattleya elongata**

ALL NAMES	ACCEPTED NAME
Cattleya aliciae	**Cattleya mossiae**
Cattleya alutacea var. *velutina*	**Cattleya velutina**
Cattleya amabilis	**Cattleya intermedia**
Cattleya amethystina	**Cattleya intermedia**
Cattleya amethystoglossa	
Cattleya anceps	**Laelia anceps**
Cattleya aquinii	**Cattleya intermedia**
Cattleya araguaiensis	
Cattleya arembergii	**Cattleya loddigesii**
Cattleya aurantiaca	
Cattleya aurea	
Cattleya autumnalis	**Cattleya bowringiana**
Cattleya autumnalis	**Laelia autumnalis***
Cattleya bassetii	**Cattleya lueddemanniana**
Cattleya batalinii	**Cattleya porphyroglossa**
Cattleya bicolor	
Cattleya bicolor var. *measuresiana*	**Cattleya bicolor**
Cattleya bogotensis	**Cattleya trianaei**
Cattleya bowringiana	
Cattleya brasiliensis	**Cattleya bicolor**
Cattleya brownii	**Cattleya harrisoniana**
Cattleya brysiana	**Laelia purpurata**
Cattleya bulbosa	**Cattleya walkeriana**
Cattleya caucaensis	**Cattleya candida**
Cattleya candida	
Cattleya carrierei	**Cattleya mossiae**
Cattleya chrysotoxa	**Cattleya aurea**
Cattleya chocoensis	**Cattleya candida**
Cattleya cinnabarina	**Laelia cinnabarina**
Cattleya coccinea	**Sophronitis coccinea**
Cattleya crispa	**Laelia crispa**
Cattleya crispa var. *purpurata*	**Laelia purpurata**
Cattleya crocata	**Cattleya eldorado**
Cattleya cupidon	**Cattleya mendelii**
Cattleya dawsonii	**Cattleya lueddemanniana**
Cattleya deckeri	
Cattleya dijanceana	**Cattleya porphyroglossa**
Cattleya dolosa	
Cattleya dormaniana	
Cattleya dowiana	
Cattleya dowiana var. *aurea*	**Cattleya aurea**
Cattleya dowiana var. *chrysotaxa*	**Cattleya aurea**
Cattleya edithiana	**Cattleya mossiae**
Cattleya elatior	**Cattleya guttata**
Cattleya eldorado	
Cattleya elegans	**Laelia elegans**
Cattleya elongata	
Cattleya epidendroides	**Cattleya luteola**
Cattleya eximia	**Cattleya dolosa**
Cattleya flava	**Laelia flava**
Cattleya flavida	**Cattleya luteola**
Cattleya forbesii	
Cattleya fragrans	**Cattleya velutina**
Cattleya fulva	**Cattleya forbesii**
Cattleya furfuracea	**Laelia furfuracea**

*For explanation see page 3, point 6

ALL NAMES	ACCEPTED NAME
Cattleya gardneriana	**Cattleya walkeriana**
Cattleya gaskelliana	
Cattleya gigas	**Cattleya warscewiczii**
Cattleya gigas var. *imperialis*	**Cattleya warscewiczii**
Cattleya gloriosa	**Cattleya warscewiczii**
Cattleya grahamii	**Laelia speciosa**
Cattleya grandiflora	**Sophronitis coccinea**
Cattleya granulosa	
Cattleya granulosa var. *dijanceana*	**Cattleya porphyroglossa**
Cattleya granulosa var. *schofeldiana*	**Cattleya schofeldiana**
Cattleya grossii	**Cattleya bicolor**
Cattleya grosvenori	**Laelia sincorana**
Cattleya guatemalensis	
Cattleya guttata	
Cattleya guttata var. *keteleerii*	**Cattleya amethystoglossa**
Cattleya guttata var. *leopoldii*	**Cattleya leopoldii**
Cattleya guttata var. *lilacina*	**Cattleya amethystoglossa**
Cattleya guttata var. *prinzii*	**Cattleya amethystoglossa**
Cattleya hardyana	
Cattleya harrisoniae	**Cattleya harrisoniana**
Cattleya harrisoniana	
Cattleya harrisonii	**Cattleya harrisoniana**
Cattleya holfordii	**Cattleya luteola**
Cattleya imperialis	**Cattleya warscewiczii**
Cattleya integerrima var. *angustifolia*	**Laelia perrinii**
Cattleya intermedia	
Cattleya intermedia var. *angustifolia*	**Laelia perrinii**
Cattleya intermedia var. *variegata*	**Cattleya harrisoniana**
Cattleya iricolor	
Cattleya isopetala	**Cattleya forbesii**
Cattleya jenmanii	
Cattleya kerrii	
Cattleya kimballiana	**Cattleya trianaei**
Cattleya labiata	
Cattleya labiata var. *atropurpurea*	**Cattleya mossiae**
Cattleya labiata var. *autumnalis*	**Cattleya labiata**
Cattleya labiata var. *bella*	**Cattleya mendelii**
Cattleya labiata var. *candida*	**Cattleya mossiae**
Cattleya labiata var. *dawsonii*	**Cattleya lueddemanniana**
Cattleya labiata var. *dowiana*	**Cattleya dowiana**
Cattleya labiata var. *eldorado*	**Cattleya eldorado**
Cattleya labiata var. *gaskelliana*	**Cattleya gaskelliana**
Cattleya labiata var. *genuina*	**Cattleya labiata**
Cattleya labiata var. *lueddemanniana*	**Cattleya lueddemanniana**
Cattleya labiata var. *mendelii*	**Cattleya mendelii**
Cattleya labiata var. *mossiae*	**Cattleya mossiae**
Cattleya labiata var. *percivaliana*	**Cattleya percivaliana**
Cattleya labiata var. *picta*	**Cattleya mossiae**
Cattleya labiata var. *reineckiana*	**Cattleya mossiae**
Cattleya labiata var. *roezlii*	**Cattleya lueddemanniana**
Cattleya labiata var. *schroderae*	**Cattleya schroderae**
Cattleya labiata var. *vera*	**Cattleya labiata**
Cattleya labiata var. *warneri*	**Cattleya warneri**
Cattleya labiata var. *warocqueana*	**Cattleya labiata**
Cattleya labiata var. *warscewiczii*	**Cattleya warscewiczii**

ALL NAMES	ACCEPTED NAME
Cattleya labiata var. *wilsoniana*	**Cattleya lueddemanniana**
Cattleya lawrenceana	
Cattleya lemoniana	**Cattleya labiata**
Cattleya leopoldii	
Cattleya lobata	**Laelia lobata**
Cattleya loddigesii	
Cattleya lueddemanniana	
Cattleya luteola	
Cattleya macmorlandii	**Cattleya eldorado**
Cattleya majalis	**Laelia speciosa**
Cattleya malouana	**Cattleya lueddemanniana***
Cattleya malouana	**Cattleya maxima**
Cattleya marginata	**Laelia pumila**
Cattleya maritima	**Cattleya intermedia**
Cattleya maxima	
Cattleya mendelii	
Cattleya meyeri	**Cattleya luteola**
Cattleya modesta	**Cattleya luteola**
Cattleya mooreana	
Cattleya mossiae	
Cattleya mossiae var. *autumnalis*	**Cattleya lueddemanniana**
Cattleya nobilior	
Cattleya odoratissima	**Cattleya violacea**
Cattleya ovata	**Cattleya intermedia**
Cattleya pachecoi	**Cattleya guatemalensis**
Cattleya papeiansiana	**Cattleya harrisoniana**
Cattleya patinii	**Cattleya deckeri**
Cattleya pauper	**Cattleya forbesii**
Cattleya peduncularis	**Laelia rubescens**
Cattleya percivaliana	
Cattleya perrinii	**Laelia perrinii**
Cattleya pinellii	**Laelia pumila**
Cattleya pinellii var. *marginata*	**Laelia pumila**
Cattleya porphyroglossa	
Cattleya princeps	**Cattleya walkeriana**
Cattleya pterocarpa	**Sophronitis pterocarpa**
Cattleya pumila	**Cattleya lawrenceana**
Cattleya pumila	**Laelia pumila***
Cattleya pumila var. *major*	**Laelia spectabilis**
Cattleya purpurata	**Laelia purpurata**
Cattleya purpurina	**Cattleya amethystoglossa**
Cattleya quadricolor	**Cattleya candida**
Cattleya quadricolor var. *eldorado*	**Cattleya eldorado**
Cattleya reflexa	**Laelia crispa**
Cattleya regnellii	**Cattleya schilleriana**
Cattleya reineckiana	**Cattleya mossiae**
Cattleya reineckiana var. *superbissima*	**Cattleya mossiae**
Cattleya rex	
Cattleya roezlii	**Cattleya lueddemanniana**
Cattleya rubescens	**Laelia rubescens**
Cattleya sanderiana	**Cattleya warscewiczii**
Cattleya schilleriana	
Cattleya schofeldiana	
Cattleya schomburgkii	**Cattleya violacea**
Cattleya schroderae	

*For explanation see page 3, point 6

ALL NAMES	ACCEPTED NAME
Cattleya schroederiana	**Cattleya walkeriana**
Cattleya skinneri	
Cattleya skinneri var. *autumnalis*	**Cattleya deckeri**
Cattleya skinneri var. *bowringiana*	**Cattleya bowringiana**
Cattleya skinneri var. *parviflora*	**Cattleya deckeri**
Cattleya skinneri var. *patinii*	**Cattleya deckeri**
Cattleya speciosissima	**Cattleya lueddemanniana**
Cattleya speciosissima var. *buchananiana*	**Cattleya lueddemanniana**
Cattleya speciosissima var. *lowii*	**Cattleya lueddemanniana**
Cattleya spectabilis	**Laelia spectabilis**
Cattleya sphenophora	**Cattleya guttata**
Cattleya sulphurea	**Cattleya luteola**
Cattleya superba	**Cattleya violacea**
Cattleya tenuis	
Cattleya tetraploidea	**Cattleya bicolor**
Cattleya tigrina	**Cattleya guttata**
Cattleya trianaei	
Cattleya trianaei var. *schroederae*	**Cattleya schroderae**
Cattleya trichopiliochila	**Cattleya eldorado**
Cattleya trilabiata	**Cattleya warneri**
Cattleya velutina	
Cattleya vestalis	**Cattleya forbesii**
Cattleya violacea*	
Cattleya violacea	**Sophronitella violacea**
Cattleya violaceum	**Cattleya violacea**
Cattleya virginalis	**Cattleya eldorado**
Cattleya wageneri	**Cattleya mossiae**
Cattleya walkeriana	
Cattleya walkeriana var. *dolosa*	**Cattleya dolosa**
Cattleya walkeriana var. *nobilior*	**Cattleya nobilior**
Cattleya wallisii	**Cattleya eldorado**
Cattleya warneri	
Cattleya warocqueana	**Cattleya labiata**
Cattleya warscewiczii	
Coelogyne arthuriana	**Pleione maculata**
Coelogyne birmanica	**Pleione praecox**
Coelogyne bulbocodioides	**Pleione bulbocodioides**
Coelogyne delavayi	**Pleione bulbocodioides**
Coelogyne diphylla	**Pleione maculata**
Coelogyne grandiflora	**Pleione grandiflora**
Coelogyne henryi	**Pleione speciosa**
Coelogyne hookeriana	**Pleione hookeriana**
Coelogyne hookeriana var. *brachyglossa*	**Pleione hookeriana**
Coelogyne humilis	**Pleione humilis**
Coelogyne humilis var. *albata*	**Pleione humilis**
Coelogyne humilis var. *tricolor*	**Pleione humilis**
Coelogyne lagenaria	**Pleione × lagenaria**
Coelogyne maculata	**Pleione maculata**
Coelogyne pogonioides	**Pleione speciosa**
Coelogyne praecox	**Pleione praecox**
Coelogyne praecox var. *sanguinea*	**Pleione praecox**
Coelogyne praecox var. *tenera*	**Pleione praecox**
Coelogyne praecox var. *wallichiana*	**Pleione praecox**
Coelogyne reichenbachiana	**Pleione praecox**
Coelogyne wallichiana	**Pleione praecox**

***For explanation see page 3, point 6**

ALL NAMES	ACCEPTED NAME
Coelogyne wallichii	**Pleione praecox**
Coelogyne yunnanensis	**Pleione yunnanensis**
Constantia australis	
Constantia cipoensis	
Constantia cristinae	
Constantia microscopica	
Constantia rupestris	
Cordula amabilis	**Paphiopedilum bullenianum**
Cordula appletoniana	**Paphiopedilum appletonianum**
Cordula argus	**Paphiopedilum argus**
Cordula barbata	**Paphiopedilum barbatum**
Cordula bellatula	**Paphiopedilum bellatulum**
Cordula boxalii	**Paphiopedilum villosum** var. **boxallii**
Cordula bulleniana	**Paphiopedilum bullenianum**
Cordula callosa	**Paphiopedilum callosum**
Cordula charlesworthii	**Paphiopedilum charlesworthii**
Cordula ciliolaris	**Paphiopedilum ciliolare**
Cordula concolor	**Paphiopedilum concolor**
Cordula curtisii	**Paphiopedilum superbiens** var. **curtisii**
Cordula dayana	**Paphiopedilum dayanum**
Cordula druryi	**Paphiopedilum druryi**
Cordula fairrieana	**Paphiopedilum fairrieanum**
Cordula glandulifera	**Paphiopedilum glanduliferum**
Cordula glaucophylla	**Paphiopedilum glaucophyllum**
Cordula godefroyae	**Paphiopedilum godefroyae**
Cordula haynaldiana	**Paphiopedilum haynaldianum**
Cordula hirsutissimum	**Paphiopedilum hirsutissimum**
Cordula hookerae	**Paphiopedilum hookerae**
Cordula insignis	**Paphiopedilum insigne**
Cordula javanica	**Paphiopedilum javanicum**
Cordula lawrenceana	**Paphiopedilum lawrenceanum**
Cordula lowii	**Paphiopedilum lowii**
Cordula mastersiana	**Paphiopedilum mastersianum**
Cordula nigrita	**Paphiopedilum barbatum**
Cordula nivea	**Paphiopedilum niveum**
Cordula parishii	**Paphiopedilum parishii**
Cordula petri	**Paphiopedilum dayanum**
Cordula philippinensis	**Paphiopedilum philippinense**
Cordula purpurata	**Paphiopedilum purpuratum**
Cordula rothschildiana	**Paphiopedilum rothschildianum**
Cordula sanderiana	**Paphiopedilum sanderianum**
Cordula spiceriana	**Paphiopedilum spicerianum**
Cordula stonei	**Paphiopedilum stonei**
Cordula superbiens	**Paphiopedilum superbiens**
Cordula tonsa	**Paphiopedilum tonsum**
Cordula venusta	**Paphiopedilum venustum**
Cordula victoria-mariae	**Paphiopedilum victoria-mariae**
Cordula villosa	**Paphiopedilum villosum**
Cordula violascens	**Paphiopedilum violascens**
Criosanthes arietina	**Cypripedium arietinum**
Criosanthes borealis	**Cypripedium arietinum**
Criosanthes parviflora	**Cypripedium arietinum**
Cymbidium amabile	**Phalaenopsis amabilis**
Cymbidium crispatum	**Laelia crispata**

ALL NAMES	ACCEPTED NAME
Cymbidium candidum	**Cattleya candida**
Cypripedium acaule	
Cypripedium acaule var. *album*	**Cypripedium acaule**
Cypripedium album	**Cypripedium reginae**
Cypripedium alternifolius	**Cypripedium calceolus**
Cypripedium amesianum	**Cypripedium tibeticum**
Cypripedium × andrewsii	
Cypripedium appletonianum	**Paphiopedilum appletonianum**
Cypripedium argus	**Paphiopedilum argus**
Cypripedium arietinum	
Cypripedium arietinum	**Cypripedium plectrochilum**
Cypripedium atsmori	**Cypripedium calceolus**
Cypripedium barbatum	**Paphiopedilum barbatum**
Cypripedium barbatum var. *biflorum*	**Paphiopedilum barbatum**
Cypripedium barbatum var. *crossii*	**Paphiopedilum callosum**
Cypripedium barbatum var. *superbum*	**Paphiopedilum superbiens**
Cypripedium barbatum var. *veitchii*	**Paphiopedilum superbiens**
Cypripedium barbatum var. *warneri*	**Paphiopedilum callosum** var. **sublaeve**
Cypripedium barbatum var. *warnerianum*	**Paphiopedilum callosum** var. **sublaeve**
Cypripedium × barbeyi	
Cypripedium bardolphianum	
Cypripedium bardolphianum var. **zhongdianense**	
Cypripedium bellatulum	**Paphiopedilum bellatulum**
Cypripedium bifidum	**Cypripedium parviflorum**
Cypripedium biflorum	**Paphiopedilum barbatum**
Cypripedium binoti	**Phragmipedium vittatum**
Cypripedium boreale	**Cypripedium calceolus**
Cypripedium boxallii	**Paphiopedilum villosum** var. **boxallii**
Cypripedium boxallii var. *atratum*	**Paphiopedilum villosum** var. **boxallii**
Cypripedium bulbosum var. *parviflorum*	**Cypripedium parviflorum**
Cypripedium bullenianum	**Paphiopedilum bullenianum**
Cypripedium bullenianum var. *appletonianum*	**Paphiopedilum appletonianum**
Cypripedium burbidgei	**Paphiopedilum dayanum**
Cypripedium calceolus	
Cypripedium calceolus var. *delta*	**Cypripedium guttatum**
Cypripedium calceolus var. *gamma*	**Cypripedium reginae**
Cypripedium californicum	
Cypripedium callosum	**Paphiopedilum callosum**
Cypripedium callosum var. *sublaeve*	**Paphiopedilum callosum** var. **sublaeve**
Cypripedium canadense	**Cypripedium reginae**
Cypripedium candidum	
Cypripedium cannartianum	**Paphiopedilum philippinense**
Cypripedium cardiophyllum	**Cypripedium debile**
Cypripedium caricinum	**Phragmipedium caricinum***
Cypripedium caricinum	**Phragmipedium pearcei**
Cypripedium cathayanum	**Cypripedium japonicum**
Cypripedium caudatum	**Phragmipedium caudatum**
Cypripedium caudatum var. *lindenii*	**Phragmipedium lindenii**
Cypripedium caudatum var. *wallisii*	**Phragmipedium wallisii**
Cypripedium chamberlainianum	**Paphiopedilum victoria-regina**
Cypripedium charlesworthii	**Paphiopedilum charlesworthii**
Cypripedium chinense	**Cypripedium henryi**

***For explanation see page 3, point 6**

ALL NAMES	ACCEPTED NAME
Cypripedium ciliolare	**Paphiopedilum ciliolare**
Cypripedium ciliolare var. *miteauanum*	**Paphiopedilum ciliolare**
Cypripedium × columbianum	
Cypripedium compactum	**Cypripedium tibeticum**
Cypripedium concolor	**Paphiopedilum concolor**
Cypripedium concolor var. *godefroyae*	**Paphiopedilum godefroyae**
Cypripedium cordigerum	
Cypripedium corrugatum	**Cypripedium tibeticum**
Cypripedium crossii	**Paphiopedilum callosum**
Cypripedium cruciforme	**Paphiopedilum lowii**
Cypripedium cruriatum	**Cypripedium calceolus**
Cypripedium curtisii	**Paphiopedilum superbiens** var. **curtisii**
Cypripedium daliense	**Cypripedium margaritaceum**
Cypripedium daultonii	**Cypripedium kentuckiense**
Cypripedium dayanum	**Paphiopedilum dayanum**
Cypripedium dayi	**Paphiopedilum dayanum**
Cypripedium debile	
Cypripedium delenatii	**Paphiopedilum delenatii**
Cypripedium dickinsonianum	
Cypripedium dilectum	**Paphiopedilum villosum** var. **boxallii**
Cypripedium druryi	**Paphiopedilum druryi**
Cypripedium ebracteatum	**Cypripedium fargesii**
Cypripedium elegans	
Cypripedium elliottianum	**Paphiopedilum rothschildianum**
Cypripedium ernestianum	**Paphiopedilum dayanum**
Cypripedium exul	**Paphiopedilum exul**
Cypripedium fairrieanum	**Paphiopedilum fairrieanum**
Cypripedium fargesii	
Cypripedium farreri	
Cypripedium fasciculatum	
Cypripedium fasciculatum var. *pusillum*	**Cypripedium fasciculatum**
Cypripedium fasciolatum	
Cypripedium × favillianum	**Cypripedium × andrewsii**
Cypripedium ferrugineum	**Cypripedium calceolus**
Cypripedium flavum	
Cypripedium formosanum	
Cypripedium forrestii	
Cypripedium franchetii	
Cypripedium × freynii	**Cypripedium × barbeyi**
Cypripedium gardineri	**Paphiopedilum glanduliferum**
Cypripedium glanduliferum	**Paphiopedilum glanduliferum**
Cypripedium glaucophyllum	**Paphiopedilum glaucophyllum**
Cypripedium godefroyae	**Paphiopedilum godefroyae**
Cypripedium godefroyae var. *leucochilum*	**Paphiopedilum godefroyae** var. **leucochilum**
Cypripedium grandiflorum	**Phragmipedium boissierianum**
Cypripedium gratrixianum	**Paphiopedilum gratrixianum**
Cypripedium guttatum	
Cypripedium guttatum subsp. *yatabeanum*	**Cypripedium yatabeanum**
Cypripedium guttatum var. *redowsky*	**Cypripedium guttatum**
Cypripedium guttatum var. *segawai*	**Cypripedium segawai**
Cypripedium guttatum var. *yatabeanum*	**Cypripedium yatabeanum**
Cypripedium haynaldianum	**Paphiopedilum haynaldianum**
Cypripedium henryi	

ALL NAMES	ACCEPTED NAME
Cypripedium himalaicum	
Cypripedium hincksianum	**Phragmipedium longifolium**
Cypripedium hirsutissimum	**Paphiopedilum hirsutissimum**
Cypripedium hirsutum	**Cypripedium acaule**
Cypripedium hirsutum var. *parviflorum*	**Cypripedium parviflorum**
Cypripedium hookerae	**Paphiopedilum hookerae**
Cypripedium hookerae var. *amabile*	**Paphiopedilum bullenianum**
Cypripedium hookerae var. *bullenianum*	**Paphiopedilum bullenianum**
Cypripedium hookerae var. *volonteanum*	**Paphiopedilum hookerae** var. **volonteanum**
Cypripedium humboldtii	**Phragmipedium caudatum**
Cypripedium humile	**Cypripedium acaule**
Cypripedium insigne	**Paphiopedilum insigne**
Cypripedium insigne var. *exul*	**Paphiopedilum exul**
Cypripedium irapeanum	
Cypripedium japonicum	
Cypripedium japonicum var. *formosanum*	**Cypripedium formosanum**
Cypripedium javanicum	**Paphiopedilum javanicum**
Cypripedium javanicum var. *virens*	**Paphiopedilum javanicum** var. **virens**
Cypripedium kentuckiense	
Cypripedium × kesselringi	**Cypripedium × barbeyi**
Cypripedium klotzschianum	**Phragmipedium klotzschianum**
Cypripedium knightae	**Cypripedium fasciculatum**
Cypripedium × krylowi	**Cypripedium × barbeyi**
Cypripedium laevigatum	**Paphiopedilum philippinense**
Cypripedium lawrenceanum	**Paphiopedilum lawrenceanum**
Cypripedium lexarzae	**Cypripedium irapeanum**
Cypripedium lichiangense	
Cypripedium lindenii	**Phragmipedium lindenii**
Cypripedium lindleyanum	**Phragmipedium lindleyanum**
Cypripedium longifolium	**Phragmipedium longifolium**
Cypripedium lowii	**Paphiopedilum lowii**
Cypripedium luteum	**Cypripedium flavum**
Cypripedium luteum	**Cypripedium parviflorum***
Cypripedium luteum var. *parviflorum*	**Cypripedium parviflorum**
Cypripedium luzmarianum	**Cypripedium irapeanum**
Cypripedium macranthon	
Cypripedium macranthon var. **rebunense**	
Cypripedium macranthum	**Cypripedium himalaicum**
Cypripedium macranthum var. *himalaicum*	**Cypripedium himalaicum**
Cypripedium macranthum var. *ventricosum*	**Cypripedium macranthon**
Cypripedium manchuricum	**Cypripedium × barbeyi**
Cypripedium margaritaceum	
Cypripedium margaritaceum var. *fargesii*	**Cypripedium fargesii**
Cypripedium marianus	**Cypripedium calceolus**
Cypripedium mastersianum	**Paphiopedilum mastersianum**
Cypripedium micranthum	
Cypripedium microsaccos	**Cypripedium calceolus**
Cypripedium miteauanum	**Paphiopedilum ciliolare**
Cypripedium molle	
Cypripedium montanum	
Cypripedium neo-guineense	**Paphiopedilum rothschildianum**
Cypripedium nigritum	**Paphiopedilum barbatum**
Cypripedium niveum	**Paphiopedilum niveum**

*For explanation see page 3, point 6

ALL NAMES	ACCEPTED NAME
Cypripedium nutans	**Cypripedium bardolphianum**
Cypripedium occidentale	**Cypripedium montanum**
Cypripedium orientale	**Cypripedium guttatum**
Cypripedium palangshanense	
Cypripedium papuanum	**Paphiopedilum papuanum**
Cypripedium pardinum	**Paphiopedilum venustum**
Cypripedium parishii	**Paphiopedilum parishii**
Cypripedium parviflorum	
Cypripedium parviflorum var. *makasin*	**Cypripedium parviflorum**
Cypripedium parviflorum var. **pubescens**	
Cypripedium passerinum	
Cypripedium passerinum var. *minganense*	**Cypripedium passerinum**
Cypripedium paulistanum	**Phragmipedium vittatum**
Cypripedium peteri	**Paphiopedilum dayanum**
Cypripedium petri	**Paphiopedilum dayanum**
Cypripedium × *petri*	**Paphiopedilum dayanum**
Cypripedium × *petri* var. *burbidgei*	**Paphiopedilum dayanum**
Cypripedium philippinense	**Paphiopedilum philippinense**
Cypripedium philippinense var. *roebelenii*	**Paphiopedilum philippinense** var. **roebelenii**
Cypripedium pitcherianum	**Paphiopedilum argus**
Cypripedium plectrochilum	
Cypripedium poyntzianum	**Paphiopedilum appletonianum**
Cypripedium praestans	**Paphiopedilum glanduliferum**
Cypripedium praestans var. *kimballianum*	**Paphiopedilum glanduliferum**
Cypripedium pubescens	**Cypripedium parviflorum**
Cypripedium pubescens var. *makasin*	**Cypripedium parviflorum**
Cypripedium pulchrum	**Cypripedium smithii**
Cypripedium purpuratum	**Paphiopedilum purpuratum**
Cypripedium pusillum	**Cypripedium fasciculatum**
Cypripedium reginae	
Cypripedium reginae var. *album*	**Cypripedium reginae**
Cypripedium robinsonii	**Paphiopedilum bullenianum**
Cypripedium roebelinii var. *cannartianum*	**Paphiopedilum philippinense**
Cypripedium roebelenii	**Paphiopedilum philippinense** var. **roebelenii**
Cypripedium rothschildianum	**Paphiopedilum rothschildianum**
Cypripedium sanderianum	**Paphiopedilum sanderianum**
Cypripedium sargentianum	**Phragmipedium sargentianum**
Cypripedium schlimii	**Phragmipedium schlimii**
Cypripedium schmidtianum	**Paphiopedilum callosum**
Cypripedium schomburgkianum	**Phragmipedium klotzschianum**
Cypripedium segawai	
Cypripedium shanxiense	
Cypripedium sinicum	**Paphiopedilum purpuratum**
Cypripedium smithii	
Cypripedium speciosum	**Cypripedium macranthon**
Cypripedium spectabile	**Cypripedium reginae**
Cypripedium spectabile var. *dayanum*	**Paphiopedilum dayanum**
Cypripedium spicerianum	**Paphiopedilum spicerianum**
Cypripedium splendidum	**Cypripedium irapeanum**
Cypripedium stonei	**Paphiopedilum stonei**
Cypripedium subtropicum	
Cypripedium superbiens	**Paphiopedilum superbiens**
Cypripedium superbiens var. *dayanum*	**Paphiopedilum dayanum**

ALL NAMES	ACCEPTED NAME
Cypripedium taiwanianum	**Cypripedium macranthon**
Cypripedium tibeticum	
Cypripedium tonkinense	**Paphiopedilum concolor**
Cypripedium tonsum	**Paphiopedilum tonsum**
Cypripedium turgidum	**Cypripedium irapeanum**
Cypripedium variegatum	**Cypripedium guttatum**
Cypripedium veitchianum	**Paphiopedilum superbiens**
Cypripedium ventricosum	**Cypripedium macranthon**
Cypripedium venustum	**Paphiopedilum venustum**
Cypripedium victoria-mariae	**Paphiopedilum victoria-mariae**
Cypripedium victoria-regina	**Paphiopedilum victoria-regina**
Cypripedium villosum	**Paphiopedilum villosum**
Cypripedium villosum var. *boxallii*	**Paphiopedilum villosum** var. **boxallii**
Cypripedium virens	**Paphiopedilum javanicum** var. **virens**
Cypripedium vittatum	**Phragmipedium vittatum**
Cypripedium volonteanum	**Paphiopedilum hookerae** var. **volonteanum**
Cypripedium waltersianum	**Paphiopedilum appletonianum**
Cypripedium wardii	
Cypripedium wardii	**Paphiopedilum wardii***
Cypripedium wilsonii	**Cypripedium fasciolatum**
Cypripedium wolterianum	**Paphiopedilum appletonianum**
Cypripedium wumengense	
Cypripedium yatabeanum	
Cypripedium yunnanense	
Cypripedium zhongdianense	**Cypripedium bardolphianum** var. **zhongdianense**
Epidendrum aclandiae	**Cattleya aclandiae**
Epidendrum amabile	**Phalaenopsis amabilis**
Epidendrum amethystoglossum	**Cattleya amethystoglossa**
Epidendrum aurantiacum	**Cattleya aurantiaca**
Epidendrum aureum	**Cattleya aurantiaca**
Epidendrum bicolor	**Cattleya bicolor**
Epidendrum canaliculatum	**Cattleya loddigesii**
Epidendrum dolosum	**Cattleya dolosa**
Epidendrum elatius	**Cattleya guttata**
Epidendrum elatius var. *leopoldii*	**Cattleya leopoldii**
Epidendrum elatius var. *prinzii*	**Cattleya amethystoglossa**
Epidendrum elegans	**Cattleya guttata**
Epidendrum forbesii	**Cattleya forbesii**
Epidendrum granulosum	**Cattleya granulosa**
Epidendrum harrisonianum	**Cattleya harrisoniana**
Epidendrum huegelianum	**Cattleya skinneri**
Epidendrum humile	**Pleione humilis**
Epidendrum humile	**Sophronitis cernua***
Epidendrum intermedium	**Cattleya intermedia**
Epidendrum iridee	**Cattleya bicolor**
Epidendrum labiatum	**Cattleya labiata**
Epidendrum labiatum var. *lueddemanniana*	**Cattleya lueddemanniana**
Epidendrum labiatum var. *mossiae*	**Cattleya mossiae**
Epidendrum labiatum var. *trianaei*	**Cattleya trianaei**
Epidendrum labiatum var. *warscewiczii*	**Cattleya warscewiczii**
Epidendrum loddigesii	**Cattleya loddigesii**
Epidendrum luteolum	**Cattleya luteola**
Epidendrum maximum	**Cattleya maxima**

***For explanation see page 3, point 6**

ALL NAMES	ACCEPTED NAME
Epidendrum pauper	**Cattleya forbesii**
Epidendrum porphyroglossum	**Cattleya porphyroglossa**
Epidendrum praecox	**Pleione praecox**
Epidendrum schillerianum	**Cattleya schilleriana**
Epidendrum superbum	**Cattleya violacea**
Epidendrum violaceum	**Cattleya loddigesii**
Epidendrum violaceum	**Cattleya violacea***
Epidendrum walkerianum	**Cattleya walkeriana**
Eunannos coccinea	**Sophronitis coccinea**
Fissipes acaulis	**Cypripedium acaule**
Fissipes hirsuta	**Cypripedium acaule**
Grafia parishii	**Phalaenopsis parishii**
Gymnostylis candida	**Pleione maculata**
Hoffmannseggella bahiensis	**Laelia bahiensis**
Hoffmannseggella brevicaulis	**Laelia cowanii**
Hoffmannseggella caulescens	**Laelia caulescens**
Hoffmannseggella cinnabarina	**Laelia cinnabarina**
Hoffmannseggella crispata	**Laelia crispata**
Hoffmannseggella crispilabia	**Laelia crispilabia**
Hoffmannseggella flava	**Laelia flava**
Hoffmannseggella ghillanyi	**Laelia ghillanyi**
Hoffmannseggella harpophylla	**Laelia harpophylla**
Hoffmannseggella kautskyi	**Laelia kautskyi**
Hoffmannseggella liliputana	**Laelia liliputana**
Hoffmannseggella macrobulbosa	**Laelia gloedeniana**
Hoffmannseggella tereticaulis	**Laelia tereticaulis**
Laelia acuminata	**Laelia rubescens**
Laelia alaorii	
Laelia albida	
Laelia anceps	
Laelia anceps subsp. dawsonii	
Laelia anceps var. *barkeriana*	**Laelia anceps**
Laelia angereri	
Laelia aurantiaca	**Cattleya aurantiaca**
Laelia aurea	
Laelia autumnalis	
Laelia bahiensis	
Laelia bancalarii	
Laelia barkeriana	**Laelia anceps**
Laelia blumenscheinii	
Laelia boothiana	**Laelia lobata**
Laelia bradei	
Laelia brevicaulis	**Laelia cowanii**
Laelia briegeri	
Laelia brysiana	**Laelia elegans**
Laelia cardimii	
Laelia casperiana	**Laelia purpurata**
Laelia caulescens	
Laelia cinnabarina	
Laelia cinnabarina var. *crispilabia*	**Laelia crispilabia**
Laelia cowanii	
Laelia crispa	
Laelia crispata	
Laelia crispilabia	
Laelia dawsonii	**Laelia anceps** var. **dawsonii**

*For explanation see page 3, point 6

ALL NAMES	ACCEPTED NAME
Laelia dayana	
Laelia devoniensis	**Laelia elegans**
Laelia discolor	**Laelia albida**
Laelia dormaniana	**Cattleya dormaniana**
Laelia duveenii	
Laelia elegans	
Laelia endsfeldzii	
Laelia esalqueana	
Laelia eyermaniana	
Laelia fidelensis	
Laelia flava	
Laelia flava var. *aurantiaca*	**Laelia flava**
Laelia fulva	**Laelia flava**
Laelia furfuracea	
Laelia gardneri	
Laelia geraensis	**Laelia flava**
Laelia ghillanyi	
Laelia gigantea	**Laelia elegans**
Laelia gloedeniana	
Laelia goebeliana	**Laelia virens**
Laelia gouldiana	
Laelia gracilis	
Laelia grandiflora	**Laelia speciosa**
Laelia grandis	
Laelia grandis var. *purpurea*	**Laelia lobata**
Laelia grandis var. *tenebrosa*	**Laelia tenebrosa**
Laelia harpophylla	
Laelia harpophylla var.*dulcotensis*	**Laelia kautskyi**
Laelia hispidula	
Laelia itambana	
Laelia johniana	**Laelia virens**
Laelia jongheana	
Laelia kautskyana	**Laelia kautskyi**
Laelia kautskyi	
Laelia kettieana	
Laelia lawrenceana	**Laelia crispilabia**
Laelia liliputana	
Laelia lobata	
Laelia longipes	
Laelia longipes var. *alba*	**Laelia lucasiana**
Laelia longipes var. *fournieri*	**Laelia lucasiana**
Laelia lucasiana	
Laelia lundii	
Laelia macrobulbosa	**Laelia gloedeniana**
Laelia majalis	**Laelia speciosa**
Laelia mantiqueirae	
Laelia milleri	
Laelia mixta	
Laelia ostermayerii	**Laelia lucasiana**
Laelia pachystele	**Laelia elegans**
Laelia peduncularis	**Laelia rubescens**
Laelia perrinii	
Laelia pfisteri	
Laelia praestans	**Laelia spectabilis**
Laelia praestans var. *nobilis*	**Laelia pumila**

Part I: All Names

ALL NAMES	ACCEPTED NAME
Laelia pubescens	**Laelia rubescens**
Laelia pumila	
Laelia pumila subsp. *praestans*	**Laelia spectabilis**
Laelia pumila var. *dayana*	**Laelia dayana**
Laelia pumila var. *mirabilis*	**Laelia spectabilis**
Laelia pumila var. *praestans*	**Laelia spectabilis**
Laelia purpurata	
Laelia purpurata var. *brysiana*	**Laelia elegans**
Laelia reginae	
Laelia regnellii	**Laelia lundii**
Laelia reichenbachiana	**Laelia lundii**
Laelia rivieri	**Laelia lobata**
Laelia rubescens	
Laelia rupestris	**Laelia crispata**
Laelia sanguiloba	
Laelia sincorana	
Laelia speciosa	
Laelia spectabilis	
Laelia tenebrosa	
Laelia tereticaulis	
Laelia turneri	**Laelia elegans**
Laelia violacea	**Laelia rubescens**
Laelia virens	
Laelia wetmorei	**Laelia xanthina**
Laelia wyattiana	**Laelia purpurata**
Laelia xanthina	
Laeliocattleya dormaniana	**Cattleya dormaniana**
Laeliocattleya elegans	**Laelia elegans**
Laeliocattleya lindenii	**Laelia elegans**
Laeliocattleya pachystele	**Laelia elegans**
Laeliocattleya sayana	**Laelia elegans**
Maclenia paradoxa	**Cattleya forbesii**
Mexipedium xerophyticum	**Phragmipedium xerophyticum**
Paphiopedilum acmodontum	
Paphiopedilum adductum	
Paphiopedilum affine	**Paphiopedilum gratrixianum**
Paphiopedilum amabile	**Paphiopedilum bullenianum**
Paphiopedilum ambonensis	**Paphiopedilum bullenianum** var. **celebesense**
Paphiopedilum appletonianum	
Paphiopedilum appletonianum var. *poyntziamum*	**Paphiopedilum appletonianum**
Paphiopedilum argus	
Paphiopedilum argus var. *sriwaniae*	**Paphiopedilum argus**
Paphiopedilum armeniacum	
Paphiopedilum bacanum	**Paphiopedilum schoseri**
Paphiopedilum barbatum	
Paphiopedilum barbatum subsp. *lawrenceanum*	**Paphiopedilum lawrenceanum**
Paphiopedilum barbatum var. *argus*	**Paphiopedilum argus**
Paphiopedilum barbatum var. *nigritum*	**Paphiopedilum barbatum**
Paphiopedilum barbigerum	
Paphiopedilum bellatulum	
Paphiopedilum birkii	**Paphiopedilum callosum** var. **sublaeve**
Paphiopedilum bodegomii	**Paphiopedilum glanduliferum** var. **wilhelminae**

ALL NAMES	ACCEPTED NAME
Paphiopedilum boissierianum	**Phragmipedium boissierianum**
Paphiopedilum bougainvilleanum	
Paphiopedilum boxallii	**Paphiopedilum villosum** var. **boxallii**
Paphiopedilum braemii	**Paphiopedilum tonsum** var. **braemii**
Paphiopedilum bullenianum	
Paphiopedilum bullenianum var. **celebesense**	
Paphiopedilum burbidgei	**Paphiopedilum dayanum**
Paphiopedilum callosum	
Paphiopedilum callosum subsp. *sublaeve*	**Paphiopedilum callosum** var. **sublaeve**
Paphiopedilum callosum var. *angustipetalum*	**Paphiopedilum callosum**
Paphiopedilum callosum var. *schmidtianum*	**Paphiopedilum callosum**
Paphiopedilum callosum var. **sublaeve**	
Paphiopedilum caricinum	**Phragmipedium caricinum**
Paphiopedilum caudatum	**Phragmipedium caudatum**
Paphiopedilum caudatum var. *lindenii*	**Phragmipedium lindenii**
Paphiopedilum caudatum var. *wallisii*	**Phragmipedium wallisii**
Paphiopedilum celebesense	**Paphiopedilum bullenianum** var. **celebesense**
Paphiopedilum ceramensis	**Paphiopedilum bullenianum** var. **celebesense**
Paphiopedilum chamberlainianum	**Paphiopedilum victoria-regina**
Paphiopedilum chamberlainianum subsp. *liemianum*	**Paphiopedilum liemianum**
Paphiopedilum chamberlainianum var. *liemianum*	**Paphiopedilum liemianum**
Paphiopedilum chamberlainianum var. *primulinum*	**Paphiopedilum primulinum**
Paphiopedilum charlesworthii	
Paphiopedilum chiwuanum	**Paphiopedilum hirsutissimum**
Paphiopedilum ciliolare	
Paphiopedilum ciliolare var. *miteauanum*	**Paphiopedilum ciliolare**
Paphiopedilum concolor	
Paphiopedilum concolor var. *niveum*	**Paphiopedilum niveum**
Paphiopedilum curtisii	**Paphiopedilum superbiens** var. **curtisii**
Paphiopedilum dariense	**Phragmipedium longifolium**
Paphiopedilum dayanum	
Paphiopedilum dayanum var. *petri*	**Paphiopedilum dayanum**
Paphiopedilum delenatii	
Paphiopedilum dennisii	**Paphiopedilum wentworthianum**
Paphiopedilum devogelii	**Paphiopedilum supardii**
Paphiopedilum dianthum	
Paphiopedilum dilectum	**Paphiopedilum villosum** var. **boxallii**
Paphiopedilum dollii	**Paphiopedilum henryanum**
Paphiopedilum druryi	
Paphiopedilum elliottianum	**Paphiopedilum adductum***
Paphiopedilum elliottianum	**Paphiopedilum rothschildianum**
Paphiopedilum emersonii	
Paphiopedilum exul	
Paphiopedilum fairrieanum	
Paphiopedilum fowliei	
Paphiopedilum gardineri	**Paphiopedilum glanduliferum***
Paphiopedilum gardineri	**Paphiopedilum glanduliferum** var. **wilhelminae**
Paphiopedilum glanduliferum	
Paphiopedilum glanduliferum var. *gardineri*	**Paphiopedilum glanduliferum**
Paphiopedilum glanduliferum var. *praestans*	**Paphiopedilum glanduliferum**

*For explanation see page 3, point 6

Part I: All Names

ALL NAMES	ACCEPTED NAME
Paphiopedilum glanduliferum var. **wilhelminae**	
Paphiopedilum glaucophyllum	
Paphiopedilum glaucophyllum var. **moquetteanum**	
Paphiopedilum godefroyae	
Paphiopedilum × godefroyae	**Paphiopedilum godefroyae**
Paphiopedilum godefroyae var. **leucochilum**	
Paphiopedilum × godefroyae var. *leucochilum*	**Paphiopedilum godefroyae** var. **leucochilum**
Paphiopedilum gratrixianum	
Paphiopedilum hainanense	**Paphiopedilum appletonianum**
Paphiopedilum haynaldianum	
Paphiopedilum hennisianum	
Paphiopedilum hennisianum var. *fowliei*	**Paphiopedilum fowliei**
Paphiopedilum henryanum	
Paphiopedilum hincksianum	**Phragmipedium longifolium**
Paphiopedilum hirsutissimum	
Paphiopedilum hirsutissimum var. *chiwuanum*	**Paphiopedilum hirsutissimum**
Paphiopedilum hirsutissimum var. **esquirolei**	
Paphiopedilum hookerae	
Paphiopedilum hookerae var. *bullenianum*	**Paphiopedilum bullenianum**
Paphiopedilum hookerae var. **volonteanum**	
Paphiopedilum insigne	
Paphiopedilum insigne var. *barbigerum*	**Paphiopedilum barbigerum**
Paphiopedilum insigne var. *exul*	**Paphiopedilum exul**
Paphiopedilum javanicum	
Paphiopedilum javanicum var. **virens**	
Paphiopedilum johorense	**Paphiopedilum bullenianum**
Paphiopedilum klotzschianum	**Phragmipedium klotzschianum**
Paphiopedilum kolopakingii	
Paphiopedilum laevigatum	**Paphiopedilum philippinense**
Paphiopedilum lawrenceanum	
Paphiopedilum leucochilum	**Paphiopedilum godefroyae** var. **leucochilum**
Paphiopedilum liemianum	
Paphiopedilum liemianum var. *primulinum*	**Paphiopedilum primulinum**
Paphiopedilum lindleyanum	**Phragmipedium lindleyanum**
Paphiopedilum linii	**Paphiopedilum bullenianum**
Paphiopedilum longifolium	**Phragmipedium longifolium**
Paphiopedilum lowii	
Paphiopedilum lowii var. **richardianum**	
Paphiopedilum malipoense	
Paphiopedilum markianum	**Paphiopedilum tigrinum**
Paphiopedilum mastersianum	
Paphiopedilum micranthum	
Paphiopedilum mohrianum	
Paphiopedilum moquetteanum	**Paphiopedilum glaucophyllum** var. **moquetteanum**
Paphiopedilum nicholsonianum	**Paphiopedilum rothschildianum**
Paphiopedilum nigritum	**Paphiopedilum barbatum**
Paphiopedilum niveum	
Paphiopedilum orbum	**Paphiopedilum callosum**
Paphiopedilum papuanum	
Paphiopedilum pardinum	**Paphiopedilum venustum**
Paphiopedilum parishii	
Paphiopedilum parishii var. *dianthum*	**Paphiopedilum dianthum**

ALL NAMES	ACCEPTED NAME
Paphiopedilum petri	**Paphiopedilum dayanum**
Paphiopedilum philippinense	
Paphiopedilum philippinense var. *cannartianum*	**Paphiopedilum philippinense**
Paphiopedilum philippinense var. **roebelenii**	
Paphiopedilum praestans	**Paphiopedilum glanduliferum**
Paphiopedilum praestans subsp. *wilhelminae*	**Paphiopedilum glanduliferum** var. **wilhelminae**
Paphiopedilum praestans var. *kimballianum*	**Paphiopedilum glanduliferum**
Paphiopedilum primulinum	
Paphiopedilum primulinum var. **purpurascens**	
Paphiopedilum purpurascens	**Paphiopedilum javanicum** var. **virens**
Paphiopedilum purpuratum	
Paphiopedilum randsii	
Paphiopedilum reflexum	**Paphiopedilum callosum**
Paphiopedilum regnieri	**Paphiopedilum callosum**
Paphiopedilum richardianum	**Paphiopedilum lowii** var. **richardianum**
Paphiopedilum robinsonii	**Paphiopedilum bullenianum**
Paphiopedilum roebelenii	**Paphiopedilum philippinense** var. **roebelenii**
Paphiopedilum rothschildianum	
Paphiopedilum rothschildianum var. *elliottianum*	**Paphiopedilum rothschildianum**
Paphiopedilum sanderianum	
Paphiopedilum sangii	
Paphiopedilum schlimii	**Phragmipedium schlimii**
Paphiopedilum schoseri	
Paphiopedilum sinicum	**Paphiopedilum purpuratum**
Paphiopedilum spicerianum	
Paphiopedilum sriwaniae	**Paphiopedilum argus**
Paphiopedilum stonei	
Paphiopedilum sublaeve	**Paphiopedilum callosum** var. **sublaeve**
Paphiopedilum sukhakulii	
Paphiopedilum supardii	
Paphiopedilum superbiens	
Paphiopedilum superbiens subsp. *ciliolare*	**Paphiopedilum ciliolare**
Paphiopedilum superbiens var. **curtisii**	
Paphiopedilum thailandense	**Paphiopedilum callosum** var. **sublaeve**
Paphiopedilum tigrinum	
Paphiopedilum tonsum	
Paphiopedilum tonsum var. **braemii**	
Paphiopedilum topperi	**Paphiopedilum kolopakingii**
Paphiopedilum tortipetalum	**Paphiopedilum bullenianum**
Paphiopedilum urbanianum	
Paphiopedilum venustum	
Paphiopedilum venustum var. *pardinum*	**Paphiopedilum venustum**
Paphiopedilum 'victoria'	**Paphiopedilum supardii**
Paphiopedilum victoria-mariae	
Paphiopedilum victoria-regina	
Paphiopedilum victoria-regina subsp. *glaucophyllum*	**Paphiopedilum glaucophyllum** var. **moquetteanum**
Paphiopedilum victoria-reginae subsp. *glaucophyllum*	**Paphiopedilum glaucophyllum**
Paphiopedilum victoria-reginae subsp. *liemianum*	**Paphiopedilum liemianum**
Paphiopedilum victoria-reginae var. *primulinum*	**Paphiopedilum primulinum**

Part I: All Names

ALL NAMES	ACCEPTED NAME
Paphiopedilum villosum	
Paphiopedilum villosum var. *affine*	**Paphiopedilum gratrixianum**
Paphiopedilum villosum var. **annamense**	
Paphiopedilum villosum var. **boxallii**	
Paphiopedilum villosum var. *gratrixianum*	**Paphiopedilum gratrixianum**
Paphiopedilum violascens	
Paphiopedilum violascens var. *gautierense*	**Paphiopedilum violascens**
Paphiopedilum virens	**Paphiopedilum javanicum** var. **virens**
Paphiopedilum vittatum	**Phragmipedium vittatum**
Paphiopedilum volonteanum	**Paphiopedilum hookerae** var. **volonteanum**
Paphiopedilum wardii	
Paphiopedilum × *wardii*	**Paphiopedilum wardii**
Paphiopedilum wentworthianum	
Paphiopedilum wilhelminae	**Paphiopedilum glanduliferum** var. **wilhelminae**
Paphiopedilum wolterianum	**Paphiopedilum appletonianum**
Paphiopedilum zieckianum	**Paphiopedilum papuanum**
Paraphalaenopsis denevei	
Paraphalaenopsis labukensis	
Paraphalaenopsis laycockii	
Paraphalaenopsis serpentilingua	
Paraphalaenopsis × **thorntonii**	
Phalaenopsis acutifolia	**Phalaenopsis sumatrana**
Phalaenopsis alcicornis	**Phalaenopsis sanderiana** var. **alba**
Phalaenopsis amabilis*	
Phalaenopsis amabilis	**Phalaenopsis aphrodite**
Phalaenopsis amabilis var. *ambigua*	**Phalaenopsis aphrodite**
Phalaenopsis amabilis var. *aphrodite*	**Phalaenopsis aphrodite**
Phalaenopsis amabilis var. *aurea*	**Phalaenopsis amabilis**
Phalaenopsis amabilis var. *cinerascens*	**Phalaenopsis amabilis** var. **moluccana**
Phalaenopsis amabilis var. *dayana*	**Phalaenopsis aphrodite**
Phalaenopsis amabilis var. *erubescens*	**Phalaenopsis aphrodite**
Phalaenopsis amabilis var. *formosa*	**Phalaenopsis aphrodite**
Phalaenopsis amabilis var. *fournieri*	**Phalaenopsis amabilis**
Phalaenopsis amabilis var. *fuscata*	**Phalaenopsis aphrodite**
Phalaenopsis amabilis var. *'gloriosa'*	**Phalaenopsis amabilis**
Phalaenopsis amabilis var. *gracillima*	**Phalaenopsis amabilis**
Phalaenopsis amabilis var. *grandiflora*	**Phalaenopsis amabilis**
Phalaenopsis amabilis var. *longifolia*	**Phalaenopsis aphrodite**
Phalaenopsis amabilis var. **moluccana**	
Phalaenopsis amabilis var. *papuana*	**Phalaenopsis rosenstromii**
Phalaenopsis amabilis var. *ramosa*	**Phalaenopsis amabilis**
Phalaenopsis amabilis var. *rimestadiana*	**Phalaenopsis amabilis**
Phalaenopsis amabilis var. *rimestadiana-alba*	**Phalaenopsis amabilis**
Phalaenopsis amabilis var. *rosenstromii*	**Phalaenopsis rosenstromii**
Phalaenopsis amabilis var. *rotundifolia*	**Phalaenopsis aphrodite**
Phalaenopsis amabilis var. *ruckeri*	**Phalaenopsis amabilis** var. **moluccana**
Phalaenopsis amabilis var. *sanderiana*	**Phalaenopsis sanderiana**
Phalaenopsis ambigua	**Phalaenopsis aphrodite**
Phalaenopsis amboinensis	
Phalaenopsis aphrodite	
Phalaenopsis × *aphroditi-equestri*	**Phalaenopsis** × **intermedia**
Phalaenopsis aphrodite var. *aphrodite*	**Phalaenopsis aphrodite**
Phalaenopsis aphrodite var. *dayana*	**Phalaenopsis aphrodite**

*For explanation see page 3, point 6

ALL NAMES	ACCEPTED NAME
Phalaenopsis aphrodite var. *gloriosa*	**Phalaenopsis amabilis**
Phalaenopsis aphrodite var. *sanderae*	**Phalaenopsis sanderiana**
Phalaenopsis appendiculata	
Phalaenopsis babuyana	**Phalaenopsis aphrodite**
Phalaenopsis barrii	**Phalaenopsis tetraspis**
Phalaenopsis bastianii	
Phalaenopsis bellina	**Phalaenopsis violacea**
Phalaenopsis boxallii	**Phalaenopsis mannii**
Phalaenopsis × *brymeriana*	**Phalaenopsis** × **intermedia**
Phalaenopsis casta	**Phalaenopsis** × **leucorrhoda**
Phalaenopsis × *casta*	**Phalaenopsis** × **leucorrhoda**
Phalaenopsis × *casta* var. *superbissima*	**Phalaenopsis** × **leucorrhoda**
Phalaenopsis celebensis	
Phalaenopsis celebica	**Phalaenopsis amabilis** var. **moluccana**
Phalaenopsis cochlearis	
Phalaenopsis corningiana	
Phalaenopsis cornucervi	
Phalaenopsis cornucervi var. **picta**	
Phalaenopsis cruciata	**Phalaenopsis maculata**
Phalaenopsis cumingiana	**Phalaenopsis corningiana**
Phalaenopsis curnowiana	**Phalaenopsis schilleriana** var. **immaculata**
Phalaenopsis × *cynthia*	**Phalaenopsis** × **leucorrhoda**
Phalaenopsis delicata	**Phalaenopsis** × **intermedia**
Phalaenopsis deltonii	**Phalaenopsis bastianii**
Phalaenopsis denevei	**Paraphalaenopsis denevei**
Phalaenopsis denevei var. *alba*	**Paraphalaenopsis serpentilingua**
Phalaenopsis denisiana	**Phalaenopsis fuscata**
Phalaenopsis denticulata	**Phalaenopsis pallens** var. **denticulata**
Phalaenopsis devriesiana	**Phalaenopsis cornucervi**
Phalaenopsis × *diezii*	**Phalaenopsis** × **intermedia**
Phalaenopsis equestris	
Phalaenopsis equestris var. **alba**	
Phalaenopsis equestris var. **leucaspis**	
Phalaenopsis equestris var. **leucotanthe**	
Phalaenopsis equestris var. **rosea**	
Phalaenopsis erubescens	**Phalaenopsis aphrodite**
Phalaenopsis fasciata	
Phalaenopsis fimbriata	
Phalaenopsis fimbriata var. **sumatrana**	
Phalaenopsis fimbriata var. *tortilis*	**Phalaenopsis fimbriata**
Phalaenopsis floresensis	
Phalaenopsis foerstermannii	**Phalaenopsis pallens**
Phalaenopsis forbesii	**Phalaenopsis viridis**
Phalaenopsis formosana	**Phalaenopsis aphrodite**
Phalaenopsis formosum	**Phalaenopsis aphrodite**
Phalaenopsis fuscata	
Phalaenopsis fuscata var. *kunstleri*	**Phalaenopsis kunstleri**
Phalaenopsis × **gersenii**	
Phalaenopsis × *gertrudeae*	**Phalaenopsis** × **veitchiana**
Phalaenopsis gibbosa	
Phalaenopsis gigantea	
Phalaenopsis gigantea var. *decolorata*	**Phalaenopsis fimbriata**
Phalaenopsis gloriosa	**Phalaenopsis amabilis**
Phalaenopsis grandiflora	**Phalaenopsis amabilis**

Part I: All Names

ALL NAMES	ACCEPTED NAME
Phalaenopsis hieroglyphica	
Phalaenopsis hombronii	**Phalaenopsis amboinensis**
Phalaenopsis imperati	**Phalaenopsis speciosa** var. **imperatrix**
Phalaenopsis inscriptiosinensis	
Phalaenopsis × intermedia	
Phalaenopsis × intermedia var. *brymeriana*	**Phalaenopsis × veitchiana**
Phalaenopsis × intermedia var. *diezii*	**Phalaenopsis × intermedia**
Phalaenopsis × intermedia var. *porteana*	**Phalaenopsis × intermedia**
Phalaenopsis × intermedia var. *portei*	**Phalaenopsis × intermedia**
Phalaenopsis × intermedia var. *vesta*	**Phalaenopsis × intermedia**
Phalaenopsis javanica	
Phalaenopsis kimballiana	**Phalaenopsis reichenbachiana**
Phalaenopsis kunstleri	
Phalaenopsis labukensis	**Paraphalaenopsis labukensis**
Phalaenopsis lamelligera	
Phalaenopsis latisepala	**Phalaenopsis javanica**
Phalaenopsis laycockii	**Paraphalaenopsis laycockii**
Phalaenopsis × leucorrhoda	
Phalaenopsis × leucorrhoda var. *alba*	**Phalaenopsis × leucorrhoda**
Phalaenopsis × leucorrhoda var. *casta*	**Phalaenopsis × leucorrhoda**
Phalaenopsis × leucorrhoda var. *cynthia*	**Phalaenopsis × leucorrhoda**
Phalaenopsis × leucorrhoda var. *grandiflora*	**Phalaenopsis × leucorrhoda**
Phalaenopsis lindenii	
Phalaenopsis listeri	**Phalaenopsis lobbii**
Phalaenopsis lobbii*	
Phalaenopsis lobbii	**Phalaenopsis × intermedia**
Phalaenopsis lowii	
Phalaenopsis lueddemanniana	
Phalaenopsis lueddemanniana subsp. *pulchra*	**Phalaenopsis pulchra**
Phalaenopsis lueddemanniana var. **delicata**	
Phalaenopsis lueddemanniana var. *hieroglyphica*	**Phalaenopsis hieroglyphica**
Phalaenopsis lueddemanniana var. **ochracea**	
Phalaenopsis lueddemanniana var. *palawanensis*	**Phalaenopsis hieroglyphica**
Phalaenopsis lueddemanniana var. *pallens*	**Phalaenopsis pallens**
Phalaenopsis lueddemanniana var. *pulchra*	**Phalaenopsis pulchra**
Phalaenopsis lueddemanniana var. *purpurea*	**Phalaenopsis pulchra**
Phalaenopsis lueddemanniana var. *surigaoensis*	**Phalaenopsis hieroglyphica**
Phalaenopsis lueddemannii	**Phalaenopsis lueddemanniana**
Phalaenopsis luteola	**Phalaenopsis pantherina**
Phalaenopsis maculata	
Phalaenopsis mannii	
Phalaenopsis mariae	
Phalaenopsis mariae var. *alba*	**Phalaenopsis pallens** var. **alba**
Phalaenopsis micholitzii	
Phalaenopsis modesta	
Phalaenopsis modesta var. **bella**	
Phalaenopsis muscicola	**Phalaenopsis maculata**
Phalaenopsis mysorensis	
Phalaenopsis ochracea	**Phalaenopsis lueddemanniana** var. **ochracea**
Phalaenopsis pallens	
Phalaenopsis pallens var. **alba**	
Phalaenopsis pallens var. **denticulata**	
Phalaenopsis pallens var. *trullifera*	**Phalaenopsis pallens**
Phalaenopsis pantherina	

*For explanation see page 3, point 6

ALL NAMES	ACCEPTED NAME
Phalaenopsis parishii	
Phalaenopsis parishii var. *lobbii*	**Phalaenopsis lobbii**
Phalaenopsis paucivittata	**Phalaenopsis sumatrana** var. **paucivittata**
Phalaenopsis philippinensis	
Phalaenopsis pleihary	**Phalaenopsis amabilis**
Phalaenopsis × *portei*	**Phalaenopsis** × **intermedia**
Phalaenopsis × *porteri*	**Phalaenopsis** × **intermedia**
Phalaenopsis proboscidioides	**Phalaenopsis lowii**
Phalaenopsis psilantha	**Phalaenopsis amboinensis**
Phalaenopsis pulchra	
Phalaenopsis reichenbachiana	
Phalaenopsis rimestadiana	**Phalaenopsis amabilis**
Phalaenopsis riteiwanensis	**Phalaenopsis equestris**
Phalaenopsis robinsonii	
Phalaenopsis rosea	**Phalaenopsis equestris**
Phalaenopsis rosea var. *aurantiaca*	**Phalaenopsis equestris** var. **leucotanthe**
Phalaenopsis rosea var. *deliciosa*	**Phalaenopsis equestris**
Phalaenopsis rosea var. *leucaspis*	**Phalaenopsis equestris** var. **leucaspis**
Phalaenopsis × *rothschildiana*	**Phalaenopsis** × **leucorrhoda**
Phalaenopsis × *rothschildiana* var. *tatsuta*	**Phalaenopsis** × **leucorrhoda**
Phalaenopsis sanderiana	
Phalaenopsis sanderiana subsp. *alba*	**Phalaenopsis sanderiana** var. **alba**
Phalaenopsis sanderiana subsp. *marmorata*	**Phalaenopsis sanderiana** var. **marmorata**
Phalaenopsis sanderiana subsp. *punctata*	**Phalaenopsis sanderiana** var. **marmorata**
Phalaenopsis sanderiana var. **alba**	
Phalaenopsis sanderiana var. **marmorata**	
Phalaenopsis sanderiana var. *pulcherrima*	**Phalaenopsis sanderiana** var. **alba**
Phalaenopsis sanderiana var. *punctata*	**Phalaenopsis sanderiana** var. **marmorata**
Phalaenopsis schilleriana	
Phalaenopsis schilleriana subsp. *immaculata*	**Phalaenopsis schilleriana** var. **immaculata**
Phalaenopsis schilleriana subsp. *vestalis*	**Phalaenopsis stuartiana**
Phalaenopsis schilleriana var. *advena*	**Phalaenopsis schilleriana** var. **immaculata**
Phalaenopsis schilleriana var. *alba*	**Phalaenopsis stuartiana**
Phalaenopsis schilleriana var. *compacta-nana*	**Phalaenopsis schilleriana**
Phalaenopsis schilleriana var. *delicata*	**Phalaenopsis schilleriana**
Phalaenopsis schilleriana var. *grandiflora*	**Phalaenopsis schilleriana** var. **splendens**
Phalaenopsis schilleriana var. **immaculata**	
Phalaenopsis schilleriana var. *major*	**Phalaenopsis schilleriana** var. **splendens**
Phalaenopsis schilleriana var. *odorata*	**Phalaenopsis schilleriana**
Phalaenopsis schilleriana var. **purpurea**	
Phalaenopsis schilleriana var. **splendens**	
Phalaenopsis schilleriana var. *stuartiana*	**Phalaenopsis stuartiana**
Phalaenopsis schilleriana var. *vestalis*	**Phalaenopsis stuartiana**
Phalaenopsis schilleriana var. *viridi-maculata*	**Phalaenopsis schilleriana**
Phalaenopsis × *schillerano*	**Phalaenopsis** × **leucorrhoda**
Phalaenopsis serpentilingua	**Paraphalaenopsis serpentilingua**

ALL NAMES	ACCEPTED NAME
Phalaenopsis simonsei	**Paraphalaenopsis serpentilingua**
Phalaenopsis × singuliflora	**Phalaenopsis × gersenii**
Phalaenopsis speciosa	
Phalaenopsis speciosa subsp. *christiana*	**Phalaenopsis speciosa** var. **christiana**
Phalaenopsis speciosa subsp. *imperatrix*	**Phalaenopsis speciosa** var. **imperatrix**
Phalaenopsis speciosa var. **christiana**	
Phalaenopsis speciosa var. **imperatrix**	
Phalaenopsis speciosa var. *maculata*	**Phalaenopsis speciosa**
Phalaenopsis speciosa var. *purpurata*	**Phalaenopsis speciosa** var. **imperatrix**
Phalaenopsis speciosa var. *tetraspis*	**Phalaenopsis tetraspis**
Phalaenopsis stauroglottis	**Phalaenopsis equestris**
Phalaenopsis stobartiana	
Phalaenopsis stuartiana	
Phalaenopsis stuartiana var. *bella*	**Phalaenopsis stuartiana**
Phalaenopsis stuartiana var. *nobilis*	**Phalaenopsis stuartiana**
Phalaenopsis stuartiana var. *punctatissima*	**Phalaenopsis stuartiana**
Phalaenopsis stuartiana var. *punctulata*	**Phalaenopsis stuartiana**
Phalaenopsis sumatrana	
Phalaenopsis sumatrana subsp. *sumatrana*	**Phalaenopsis corningiana**
Phalaenopsis sumatrana var. **alba**	
Phalaenopsis sumatrana var. *gersenii*	**Phalaenopsis × gersenii**
Phalaenopsis sumatrana var. *kimballiana*	**Phalaenopsis sumatrana**
Phalaenopsis sumatrana var. **paucivittata**	
Phalaenopsis sumatrana var. *sanguinea*	**Phalaenopsis corningiana**
Phalaenopsis tetraspis	
Phalaenopsis thalebanii	
Phalaenopsis × thorntonii	**Paraphalaenopsis × thorntonii**
Phalaenopsis × valentinii	
Phalaenopsis × veitchiana	
Phalaenopsis × veitchiana var. *brachyodon*	**Phalaenopsis × veitchiana**
Phalaenopsis veitchii	**Phalaenopsis × veitchiana**
Phalaenopsis venosa	
Phalaenopsis × vesta	**Phalaenopsis × intermedia**
Phalaenopsis violacea	
Phalaenopsis violacea subsp. *alba*	**Phalaenopsis violacea** var. **alba**
Phalaenopsis violacea var. **alba**	
Phalaenopsis violacea var. *bellina*	**Phalaenopsis violacea**
Phalaenopsis violacea var. **bowringiana**	
Phalaenopsis violacea var. *chloracea*	**Phalaenopsis violacea**
Phalaenopsis violacea var. **murtoniana**	
Phalaenopsis violacea var. **punctata**	
Phalaenopsis violacea var. *schroederi*	**Phalaenopsis × gersenii**
Phalaenopsis violacea var. *schroederiana*	**Phalaenopsis × gersenii**
Phalaenopsis × virataii	**Phalaenopsis × veitchiana**
Phalaenopsis viridis	
Phalaenopsis wilsonii	
Phalaenopsis × youngiana	**Phalaenopsis × leucorrhoda**
Phalaenopsis × youngii	**Phalaenopsis × leucorrhoda**
Phalaenopsis zebrina	**Phalaenopsis sumatrana**
Phalaenopsis zebrina var. *gersenii*	**Phalaenopsis × gersenii**
Phalaenopsis zebrina var. *lilacina*	**Phalaenopsis × gersenii**
Phragmipedilum sargentianum	**Phragmipedium sargentianum**
Phragmipedium besseae	
Phragmipedium boissierianum	

ALL NAMES	ACCEPTED NAME
Phragmipedium cajamarcae	**Phragmipedium boissierianum**
Phragmipedium caricinum	
Phragmipedium caudatum	
Phragmipedium caudatum var. *lindenii*	**Phragmipedium lindenii**
Phragmipedium czerwiakowianum	**Phragmipedium boissierianum**
Phragmipedium dariense	**Phragmipedium longifolium**
Phragmipedium ecuadorense	**Phragmipedium pearcei**
Phragmipedium exstaminodium	
Phragmipedium hartwegii	**Phragmipedium longifolium**
Phragmipedium hincksianum	**Phragmipedium longifolium**
Phragmipedium hirtzii	
Phragmipedium kaieteurum	**Phragmipedium lindleyanum** var. **kaieteurum**
Phragmipedium klotzschianum	
Phragmipedium lindenii	
Phragmipedium lindleyanum	
Phragmipedium lindleyanum var. **kaieteurum**	
Phragmipedium longifolium	
Phragmipedium pearcei	
Phragmipedium portillae	
Phragmipedium reticulatum	**Phragmipedium boissierianum**
Phragmipedium richteri	
Phragmipedium roezlii	**Phragmipedium longifolium**
Phragmipedium sargentianum	
Phragmipedium schlimii	
Phragmipedium vittatum	
Phragmipedium wallisii	
Phragmipedium warszewiczianum	**Phragmipedium caudatum**
Phragmipedium xerophyticum	
Phragmopedilum sargentianum	**Phragmipedium sargentianum**
Pleione alba	**Pleione forrestii**
Pleione albiflora	
Pleione amoena	
Pleione aurita	**Pleione chunii**
Pleione birmanica	**Pleione praecox**
Pleione bulbocodioides	
Pleione chiwuana	**Pleione yunnanensis**
Pleione chunii	
Pleione communis	**Pleione bulbocodioides**
Pleione communis var. *subobtusum*	**Pleione bulbocodioides**
Pleione concolor	**Pleione praecox**
Pleione × confusa	
Pleione coronaria	
Pleione delavayi	**Pleione bulbocodioides**
Pleione diantha	**Pleione humilis**
Pleione diphylla	**Pleione maculata**
Pleione fargesii	**Pleione bulbocodioides**
Pleione formosana	
Pleione forrestii	
Pleione ganchuenensis	**Pleione bulbocodioides**
Pleione grandiflora	
Pleione henryi	**Pleione speciosa**
Pleione hookeriana	
Pleione hookeriana var. *brachyglossa*	**Pleione hookeriana**
Pleione hui	**Pleione formosana**

Part I: All Names

ALL NAMES	ACCEPTED NAME
Pleione humilis	
Pleione humilis var. *adnata*	**Pleione humilis**
Pleione humilis var. *pulchella*	**Pleione humilis**
Pleione humilis var. *purpurascens*	**Pleione humilis**
Pleione × kohlsii	
Pleione × lagenaria	
Pleione laotica	**Pleione hookeriana**
Pleione limprichtii	
Pleione maculata	
Pleione maculata var. *arthuriana*	**Pleione maculata**
Pleione maculata var. *virginea*	**Pleione maculata**
Pleione mairei	**Pleione bulbocodioides**
Pleione pinkepankii	**Pleione albiflora**
Pleione pogonioides	**Pleione bulbocodioides***
Pleione pogonioides	**Pleione speciosa**
Pleione praecox	
Pleione praecox var. *alba*	**Pleione praecox**
Pleione praecox var. *birmanica*	**Pleione praecox**
Pleione praecox var. *candida*	**Pleione praecox**
Pleione praecox var. *sanguinea*	**Pleione praecox**
Pleione praecox var. *wallichiana*	**Pleione praecox**
Pleione pricei	**Pleione formosana**
Pleione reichenbachiana	**Pleione praecox**
Pleione rhombilabia	**Pleione bulbocodioides**
Pleione saxicola	
Pleione scopulorum	
Pleione smithii	**Pleione bulbocodioides**
Pleione speciosa	
Pleione wallichiana	**Pleione praecox**
Pleione yunnanensis	
Polychilos amboinensis	**Phalaenopsis amboinensis**
Polychilos cochlearis	**Phalaenopsis cochlearis**
Polychilos corningiana	**Phalaenopsis corningiana**
Polychilos cornucervi	**Phalaenopsis cornucervi**
Polychilos fasciata	**Phalaenopsis fasciata**
Polychilos fimbriata	**Phalaenopsis fimbriata**
Polychilos fuscata	**Phalaenopsis fuscata**
Polychilos gigantea	**Phalaenopsis gigantea**
Polychilos hieroglyphica	**Phalaenopsis hieroglyphica**
Polychilos javanica	**Phalaenopsis javanica**
Polychilos kunstleri	**Phalaenopsis kunstleri**
Polychilos lobbii	**Phalaenopsis lobbii**
Polychilos lowii	**Phalaenopsis lowii**
Polychilos lueddemanniana	**Phalaenopsis lueddemanniana**
Polychilos maculata	**Phalaenopsis maculata**
Polychilos mannii	**Phalaenopsis mannii**
Polychilos mariae	**Phalaenopsis mariae**
Polychilos micholitzii	**Phalaenopsis micholitzii**
Polychilos modesta	**Phalaenopsis modesta**
Polychilos pallens	**Phalaenopsis pallens**
Polychilos pantherina	**Phalaenopsis pantherina**
Polychilos parishii	**Phalaenopsis parishii**
Polychilos pulchra	**Phalaenopsis pulchra**
Polychilos speciosa	**Phalaenopsis speciosa**
Polychilos sumatrana	**Phalaenopsis sumatrana**

*For explanation see page 3, point 6

ALL NAMES	ACCEPTED NAME
Polychilos venosa	**Phalaenopsis venosa**
Polychilos violacea	**Phalaenopsis violacea**
Polychilos viridis	**Phalaenopsis viridis**
Polystylus cornucervi	**Phalaenopsis cornucervi**
Polystylus cornucervi var. *picta*	**Phalaenopsis cornucervi** var. **picta**
Sacodon macranthum	**Cypripedium macranthon**
Selenipedium boissierianum	**Phragmipedium boissierianum**
Selenipedium caricinum	**Phragmipedium caricinum**
Selenipedium caudatum	**Phragmipedium caudatum**
Selenipedium caudatum var. *lindenii*	**Phragmipedium lindenii**
Selenipedium caudatum var. *uropedium*	**Phragmipedium lindenii**
Selenipedium caudatum var. *wallisii*	**Phragmipedium wallisii**
Selenipedium dariense	**Phragmipedium longifolium**
Selenipedium duboisii	**Phragmipedium boissierianum**
Selenipedium duboissierianum	**Phragmipedium boissierianum**
Selenipedium elliottianum	**Paphiopedilum rothschildianum**
Selenipedium kaieteurum	**Phragmipedium lindleyanum** var. **kaieteurum**
Selenipedium klotzschianum	**Phragmipedium klotzschianum**
Selenipedium laevigatum	**Paphiopedilum philippinense**
Selenipedium lindenii	**Phragmipedium lindenii**
Selenipedium lindleyanum	**Phragmipedium lindleyanum**
Selenipedium lindleyanum var. *kaieteurum*	**Phragmipedium lindleyanum** var. **kaieteurum**
Selenipedium longifolium	**Phragmipedium longifolium**
Selenipedium parishii	**Paphiopedilum parishii**
Selenipedium paulistanum	**Phragmipedium vittatum**
Selenipedium pearcei	**Phragmipedium pearcei**
Selenipedium reichenbachii	**Phragmipedium longifolium**
Selenipedium sargentianum	**Phragmipedium sargentianum**
Selenipedium schlimii	**Phragmipedium schlimii**
Selenipedium schomburgkianum	**Phragmipedium klotzschianum**
Selenipedium vittatum	**Phragmipedium vittatum**
Selenipedium wallisii	**Phragmipedium wallisii**
Sophronia cernua	**Sophronitis cernua**
Sophronitella violacea	
Sophronitis acuensis	
Sophronitis australis	**Constantia australis**
Sophronitis bicolor	
Sophronitis brevipedunculata	
Sophronitis cernua	
Sophronitis cernua var. *pterocarpa*	**Sophronitis pterocarpa**
Sophronitis coccinea	
Sophronitis coccinea subsp. *mantiqueirae*	**Sophronitis mantiqueirae**
Sophronitis coccinea subsp. *pygmaea*	**Sophronitis pygmaea**
Sophronitis coccinea var. *parviflora*	**Sophronitis mantiqueirae**
Sophronitis coccinea var. *rossiteriana*	**Sophronitis coccinea**
Sophronitis grandiflora	**Sophronitis coccinea**
Sophronitis grandiflora var. *rosea*	**Sophronitis wittigiana**
Sophronitis hoffmannseggii	**Sophronitis cernua**
Sophronitis lowii	**Sophronitis coccinea**
Sophronitis mantiqueirae	
Sophronitis militaris	**Sophronitis coccinea**
Sophronitis modesta	**Sophronitis cernua**
Sophronitis nutans	**Sophronitis cernua**

Part I: All Names

ALL NAMES	ACCEPTED NAME
Sophronitis pterocarpa	
Sophronitis purpurea	**Sophronitis wittigiana**
Sophronitis pygmaea	
Sophronitis rosea	**Sophronitis wittigiana**
Sophronitis rossiteriana	**Sophronitis coccinea**
Sophronitis rupestris	**Constantia rupestris**
Sophronitis violacea	**Sophronitella violacea**
Sophronitis violacea	**Sophronitis wittigiana***
Sophronitis wittigiana	
Sophronitis wittigiana var. *brevipedunculata*	**Sophronitis brevipedunculata**
Staurites violacea	**Phalaenopsis violacea**
Stauroglottis equestris	**Phalaenopsis equestris**
Stauroglottis riteiwanensis	**Phalaenopsis equestris**
Stauropsis pallens	**Phalaenopsis pallens**
Stauropsis violacea	**Phalaenopsis violacea**
Stimegas venustum	**Paphiopedilum venustum**
Synadena amabilis	**Phalaenopsis amabilis**
Trichoglottis pallens	**Phalaenopsis pallens**
Uropedium lindenii	**Phragmipedium lindenii**

*For explanation see page 3, point 6

PART II: ORCHIDACEAE BINOMIALS IN CURRENT USAGE
Ordered Alphabetically on Accepted Names for the Genera

Cattleya, Constantia, Cypripedium, Laelia, Paphiopedilum, Paraphalaenopis, Phalaenopsis, Phragmipedium, Pleione, Sophronitella and *Sophronitis*

CATTLEYA BINOMIALS IN CURRENT USAGE

Cattleya aclandiae Lindl.
Epidendrum aclandiae (Lindl.) Rchb.f.

Distribution: Brazil

Cattleya amethystoglossa Linden & Rchb.f. ex R.Warner
Cattleya guttata var. *keteleerii* Houllet
Cattleya guttata var. *lilacina* Rchb.f.
Cattleya guttata var. *prinzii* Rchb.f.
Cattleya purpurina Barb.Rodr.
Epidendrum amethystoglossum (Lindl. & Rchb.f.) Rchb.f.
Epidendrum elatius var. *prinzii* (Rchb.f.) Rchb.f.

Distribution: Brazil

Cattleya araguaiensis Pabst

Distribution: Brazil

Cattleya aurantiaca (Bateman ex Lindl.) P.N.Don
Broughtonia aurea Lindl.
Epidendrum aurantiacum Bateman ex Lindl.
Epidendrum aureum (Lindl.) Lindl.
Laelia aurantiaca (Lindl.) Beer

Distribution: El Salvador, Guatemala, Honduras, Mexico

Cattleya aurea (T.Moore, R.Warner & B.S.Williams) Rodigas
Cattleya chrysotoxa (Sander) God.-Lebeuf
Cattleya dowiana var. *aurea* T.Moore, R.Warner & B.S.Williams
Cattleya dowiana var. *chrysotaxa* hort. ex Sander

Distribution: Colombia

Cattleya bicolor Lindl.
Cattleya bicolor var. *measuresiana* R.Warner & B.S.Williams
Cattleya brasiliensis Klinge
Cattleya grossii Kraenzl.
Cattleya tetraploidea F.G.Brieger
Epidendrum bicolor (Lindl.) Rchb.f.
Epidendrum iridee Descourtilz

Distribution: Brazil

Cattleya bowringiana O'Brien
Cattleya autumnalis hort. ex O'Brien
Cattleya skinneri var. *bowringiana* Kraenzl.

Distribution: Belize, Guatemala

Cattleya candida (Kunth) Lehm.
Cattleya caucaensis Roezl ex Ballif.
Cattleya chocoensis Linden
Cattleya quadricolor Bateman
Cymbidium candidum Kunth

Distribution: Colombia

Cattleya deckeri Klotzsch
Cattleya patinii hort. ex Cogn.
Cattleya skinneri var. *autumnalis* Allen
Cattleya skinneri var. *parviflora* Hook.
Cattleya skinneri var. *patinii* (Cogn.) Schltr.

Distribution: Costa Rica, El Salvador, Guatemala, Honduras, Nicaragua, Colombia, Trinidad & Tobago, Venezuela

Cattleya dolosa (Rchb.f.) Rchb.f.
Cattleya eximia Barb.Rodr.
Cattleya walkeriana var. *dolosa* Veitch
Epidendrum dolosum Rchb.f.

Distribution: Brazil

Cattleya dormaniana (Rchb.f.) Rchb.f.
Laelia dormaniana Rchb.f.
Laeliocattleya dormaniana (Rchb.f.) Rolfe

Distribution: Brazil

Cattleya dowiana Bateman
Cattleya labiata var. *dowiana* (Bateman) Veitch

Distribution: Costa Rica

Cattleya eldorado Linden ex Van Houtte
Cattleya crocata Rchb.f.
Cattleya labiata var. *eldorado* (Linden ex Van Houtte) Veitch
Cattleya macmorlandii Nicholson
Cattleya quadricolor var. *eldorado* (Linden ex Van Houtte) Morren & Devos
Cattleya trichopiliochila Barb.Rodr.
Cattleya virginalis Linden & Andre ex Devos
Cattleya wallisii Linden ex Rchb.f.

Distribution: Brazil

Cattleya elongata Barb.Rodr.
Cattleya alexandrae L.Linden & Rolfe

Distribution: Brazil

Cattleya forbesii Lindl.
Cattleya fulva Beer
Cattleya isopetala Beer
Cattleya pauper (Vell.) Stellfeld
Cattleya vestalis Hoffmanns.
Epidendrum forbesii (Lindl.) Rchb.f.
Epidendrum pauper Vell.
Maclenia paradoxa DuMort.

Distribution: Brazil

Cattleya gaskelliana (hort. ex N.E.Br.) Withner
Cattleya labiata var. *gaskelliana* hort. ex N.E.Br.

Distribution: Colombia, Venezuela

Cattleya granulosa Lindl.
Epidendrum granulosum (Lindl.) Rchb.f.

Distribution: Brazil

Cattleya guatemalensis T.Moore
Cattleya pachecoi Ames & Correll

Distribution: Guatemala

Cattleya guttata Lindl.
Cattleya elatior Lindl.
Cattleya sphenophora C.Morren
Cattleya tigrina A.Rich. ex Beer
Epidendrum elatius (Lindl.) Rchb.f.
Epidendrum elegans Vell.

Distribution: Brazil

Cattleya hardyana Hardy ex B.S.Williams

Distribution: Colombia

Cattleya harrisoniana Bateman ex Lindl.
Cattleya brownii Rolfe
Cattleya harrisoniae Paxton
Cattleya harrisonii Duchartre
Cattleya intermedia var. *variegata* Hook.

Cattleya papeiansiana C.Morren
Epidendrum harrisonianum (Bateman ex Lindl.) Rchb.f.

Distribution: Brazil

Cattleya intermedia Graham ex Hook.
Cattleya amabilis hort. ex B.S.Williams
Cattleya amethystina C.Morren
Cattleya aquinii Barb.Rodr.
Cattleya maritima Lindl.
Cattleya ovata Lindl.
Epidendrum intermedium (Graham) Rchb.f.

Distribution: Brazil

Cattleya iricolor Rchb.f.

Distribution: Ecuador

Cattleya jenmanii Rolfe

Distribution: Guyana, Venezuela

Cattleya kerrii F.G.Brieger & Bicalho

Distribution: Brazil

Cattleya labiata Lindl.
Cattleya labiata var. *autumnalis* Linden
Cattleya labiata var. *genuina* Stein
Cattleya labiata var. *vera* Veitch
Cattleya labiata var. *warocqueana* Rolfe
Cattleya lemoniana Lindl.
Cattleya warocqueana Linden ex Kerchove
Epidendrum labiatum (Lindl.) Rchb.f. non Swartz

Distribution: Brazil

Cattleya lawrenceana Rchb.f.
Cattleya pumila auct. non Hook.; M.R.Schomb.

Distribution: Guyana, Venezuela

Cattleya leopoldii Verschaffelt ex Lem.
Cattleya guttata var. *leopoldii* Lem.
Epidendrum elatius var. *leopoldii* Rchb.f.

Distribution: Brazil

Cattleya loddigesii Lindl.
Cattleya arembergii Scheidw.
Epidendrum canaliculatum Vell.
Epidendrum loddigesii (Lindl.) Rchb.f.
Epidendrum violaceum Lodd.

Distribution: Argentina, Brazil

Cattleya lueddemanniana Rchb.f.
Cattleya bassetii hort. ex Veitch
Cattleya dawsonii hort. ex R.Warner
Cattleya labiata var. *dawsonii* (R.Warner) Dubuysson
Cattleya labiata var. *lueddemanniana* (Rchb.f.) Rchb.f.
Cattleya labiata var. *roezlii* Rchb.f.
Cattleya labiata var. *wilsoniana* Rchb.f.
Cattleya malouana Linden
Cattleya mossiae var. *autumnalis* (hort.) Veitch
Cattleya roezlii (Rchb.f.) Rchb.f.
Cattleya speciosissima hort.
Cattleya speciosissima var. *buchananiana* hort.
Cattleya speciosissima var. *lowii* hort.
Epidendrum labiatum var. *lueddemanniana* (Rchb.f.) Rchb.f.

Distribution: Venezuela

Cattleya luteola Lindl.
Cattleya epidendroides hort. ex Rchb.f.
Cattleya flavida Klotzsch
Cattleya holfordii hort. ex Rchb.f.
Cattleya meyeri Regel
Cattleya modesta Meyer ex Regel
Cattleya sulphurea hort.
Epidendrum luteolum (Lindl.) Rchb.f.

Distribution: Bolivia, Brazil, Ecuador, Peru

Cattleya maxima Lindl.
Cattleya malouana L.Linden ex Linden & Rodigas
Epidendrum maximum (Lindl.) Rchb.f.

Distribution: Colombia, Ecuador, Peru, Venezuela

Cattleya mendelii Backh.f. ex B.S.Williams
Cattleya cupidon hort. ex L.Linden et Rodigas
Cattleya labiata var. *bella* Rchb.f.
Cattleya labiata var. *mendelii* Sanders

Distribution: Colombia

Part II: Cattleya

Cattleya mooreana Withner, Allison & Guenard

Distribution: Peru

Cattleya mossiae Hook.
Cattleya aliciae L.Linden
Cattleya carrierei Houllet
Cattleya edithiana R.Warner ex B.S.Williams
Cattleya labiata var. *atropurpurea* Paxton
Cattleya labiata var. *candida* Lindl. & Paxton
Cattleya labiata var. *mossiae* (Hook.) Lindl.
Cattleya labiata var. *picta* Lindl. & Paxton
Cattleya labiata var. *reineckiana* Rchb.f.
Cattleya reineckiana Rchb.f.
Cattleya reineckiana var. *superbissima* Hast.
Cattleya wageneri Rchb.f.
Epidendrum labiatum var. *mossiae* (Hook.) Rchb.f.

Distribution: Venezuela

Cattleya nobilior Rchb.f.
Cattleya walkeriana var. *nobilior* Veitch

Distribution: Brazil

Cattleya percivaliana (Rchb.f.) O'Brien
Cattleya labiata var. *percivaliana* Rchb.f.

Distribution: Venezuela

Cattleya porphyroglossa Linden & Rchb.f.
Cattleya batalinii Sander et Kraenzl.
Cattleya dijanceana hort. ex Rolfe
Cattleya granulosa var. *dijanceana* Veitch
Epidendrum porphyroglossum Linden & Rchb.f.

Distribution: Brazil

Cattleya rex O'Brien

Distribution: Peru

Cattleya schilleriana Rchb.f.
Cattleya aclandiae var. *schilleriana* Jennings
Cattleya regnellii hort. ex R.Warner
Epidendrum schillerianum (Rchb.f.) Rchb.f.

Distribution: Brazil

Cattleya schofeldiana Rchb.f.
Cattleya granulosa var. *schofeldiana* (Rchb.f.) Veitch

Distribution: Brazil

Cattleya schroderae (Rchb.f.) Sander
Cattleya labiata var. *schroderae* Sander
Cattleya trianaei var. *schroederae* Rchb.f.

Distribution: Colombia

Cattleya skinneri Bateman
Epidendrum huegelianum Rchb.f.

Distribution: Costa Rica, Guatemala, Honduras, Nicaragua, Panama

Cattleya tenuis Campacci & Vedovello

Distribution: Brazil

Cattleya trianaei Linden & Rchb.f.
Cattleya bogotensis Linden
Cattleya kimballiana L.Linden & Rodigas
Epidendrum labiatum var. *trianaei* Rchb.f.

Distribution: Colombia

Cattleya velutina Rchb.f.
Cattleya alutacea var. *velutina* Barb.Rodr.
Cattleya fragrans Barb.Rodr.

Distribution: Brazil

Cattleya violacea (Kunth) Rolfe
Cattleya odoratissima P.N.Don
Cattleya schomburgkii Lodd. ex Lindl.
Cattleya superba R.H.Schomb. ex Lindl.
Cattleya violaceum Kunth
Epidendrum superbum (Schomb. ex Lindl.) Rchb.f.
Epidendrum violaceum Rchb.f.

Distribution: Brazil, Colombia, Ecuador, Guyana, Peru, Venezuela

Cattleya walkeriana Gardner
Cattleya bulbosa Lindl.
Cattleya gardneriana Rchb.f.
Cattleya princeps Barb.Rodr.
Cattleya schroederiana Rchb.f.

Epidendrum walkerianum (Gardner) Rchb.f.

Distribution: Brazil

Cattleya warneri T.Moore ex R.Warner
Cattleya labiata var. *warneri* O'Brien
Cattleya trilabiata Barb.Rodr.

Distribution: Brazil

Cattleya warscewiczii Rchb.f.
Cattleya gigas Linden ex Andre
Cattleya gigas var. *imperialis* O'Brien
Cattleya gloriosa Carriere
Cattleya imperialis hort. ex Veitch
Cattleya labiata var. *warscewiczii* Rchb.f.
Cattleya sanderiana Rchb.f.
Epidendrum labiatum var. *warscewiczii* Rchb.f.

Distribution: Colombia

CONSTANTIA BINOMIALS IN CURRENT USAGE

Constantia australis (Cogn.) Porto & Brade
Sophronitis australis Cogn.

Distribution: Brazil

Constantia cipoensis Porto & Brade

Distribution: Brazil

Constantia cristinae F.Miranda

Distribution: Brazil

Constantia microscopica F.Miranda

Distribution: Brazil

Constantia rupestris Barb. Rodr.
Sophronitis rupestris (Barb. Rodr.) Cogn.

Distribution: Brazil

CYPRIPEDIUM BINOMIALS IN CURRENT USAGE

Cypripedium acaule Aiton
Cypripedium acaule var. *album* Ballif
Cypripedium hirsutum (Farw.) Hiller
Cypripedium humile Salisb.
Fissipes acaulis (Aiton) Small
Fissipes hirsuta Farw.

Distribution: Canada, USA

Cypripedium × andrewsii Fuller
Cypripedium × favillianum Curtis

Distribution: USA

Cypripedium arietinum R.Br.
Arietinum americanum Beck
Criosanthes arietina (R.Br.) House
Criosanthes borealis Raf.
Criosanthes parviflora Raf.

Distribution: Canada, USA

Cypripedium × barbeyi Camus
Cypripedium × freynii Karo
Cypripedium × kesselringi G.Keller
Cypripedium × krylowi Sjuzew
Cypripedium manchuricum Stapf

Distribution: China, Korea, Russian Federation

Cypripedium bardolphianum W.W.Sm. & Farrer
Cypripedium nutans Schltr.

Distribution: China

Cypripedium bardolphianum var. **zhongdianense** S.C.Chen
Cypripedium zhongdianense S.C.Chen

Distribution: China

Cypripedium calceolus L.
Cypripedium alternifolius St.-Leg.
Cypripedium atsmori Morren
Cypripedium boreale E.Salisb.
Cypripedium cruriatum Dulac
Cypripedium ferrugineum Gray

Cypripedium marianus Rouy
Cypripedium marianus Crantz
Cypripedium microsaccos Kraenzl.

Distribution: Austria, Belgium, Bulgaria, Canada, China, Czech Republic, Denmark, Finland, France, Germany, Greece, Hungary, Italy, Japan, Korea, Lithuania, Luxembourg, Mongolia, Norway, Poland, Romania, Russian Federation, Slovak Republic, Spain, Sweden, Switzerland, United Kingdom, USA, Yugoslavia (former)

Cypripedium californicum A.Gray

Distribution: USA

Cypripedium candidum Muhl. ex Willd.
Calceolus candidus (Muhl. ex Willd.) Nieuwl.

Distribution: Canada, USA

Cypripedium × columbianum Sheviak

Distribution: Canada, USA

Cypripedium cordigerum D.Don

Distribution: Bhutan, India, Nepal, Pakistan, China (Tibet)

Cypripedium debile Rchb.f.
Cypripedium cardiophyllum Franch. & Sav.

Distribution: China, Japan, China (Taiwan)

Cypripedium dickinsonianum Hágsater

Distribution: Mexico

Cypripedium elegans Rchb.f.

Distribution: Bhutan, China, India, Nepal

Cypripedium fargesii Franch.
Cypripedium ebracteatum Rolfe
Cypripedium margaritaceum var. *fargesii* (Franch.) Pfitzer

Distribution: China

Cypripedium farreri W.W.Sm.

Distribution: China

Cypripedium fasciculatum Kellogg ex S.Watson
Cypripedium fasciculatum var. *pusillum* (Rolfe) Hook.f.
Cypripedium knightae A.Nelson
Cypripedium pusillum Rolfe

Distribution: Canada, USA

Cypripedium fasciolatum Franch.
Cypripedium wilsonii Rolfe

Distribution: China

Cypripedium flavum Hunt & Summerh.
Cypripedium luteum Franch.

Distribution: China, China (Tibet)

Cypripedium formosanum Hayata
Cypripedium japonicum var. *formosanum* (Hayata.) Ying

Distribution: China (Taiwan)

Cypripedium forrestii P.J.Cribb

Distribution: China

Cypripedium franchetii Rolfe

Distribution: China

Cypripedium guttatum Sw.
Cypripedium calceolus var. *delta* L.
Cypripedium guttatum var. *redowsky* Rchb.f.
Cypripedium orientale Spreng.
Cypripedium variegatum Georgi

Distribution: Bhutan, China, India, Japan, Korea, Nepal, Russian Federation, USA

Cypripedium henryi Rolfe
Cypripedium chinense Franch.

Distribution: China

Part II: Cypripedium

Cypripedium himalaicum Rolfe ex Hemsl.
Cypripedium macranthum auct. non Sw.; Hook.f.
Cypripedium macranthum var. *himalaicum* Kraenzl.

Distribution: China, Nepal, Bhutan, India

Cypripedium irapeanum La Llave & Lex.
Cypripedium lexarzae Scheidw.
Cypripedium luzmarianum R.Gonzalez & R.Ramirez
Cypripedium splendidum Scheidw.
Cypripedium turgidum Sesse & Mocino

Distribution: Guatemala, Honduras, Mexico

Cypripedium japonicum Thunb.
Cypripedium cathayanum S.S.Chien

Distribution: China, Japan, Korea

Cypripedium kentuckiense C.F.Reed
Cypripedium daultonii Soukup

Distribution: USA

Cypripedium lichiangense P.J.Cribb & S.C.Chen

Distribution: China, Myanmar

Cypripedium macranthon Sw.
Cypripedium macranthum var. *ventricosum* (Sw.) Rchb.f.
Cypripedium speciosum Rolfe
Cypripedium taiwanianum Masam.
Cypripedium ventricosum Sw.
Sacodon macranthum (Sw.) Raf.

Distribution: China, Japan, Korea, Russian Federation, China (Taiwan)

Cypripedium macranthon var. **rebunense** Miyabe & Kudo

Distribution: China, Japan, Russian Federation

Cypripedium margaritaceum Franch.
Cypripedium daliense Chen & Wu

Distribution: China

Cypripedium micranthum Franch.

Distribution: China

Cypripedium molle Lindl.

Distribution: Mexico

Cypripedium montanum Douglas ex Lindl.
Cypripedium occidentale S.Watson

Distribution: Canada, USA

Cypripedium palangshanense Tang & Wang

Distribution: China

Cypripedium parviflorum E.Salisb.
Cypripedium bifidum Raf.
Cypripedium bulbosum var. *parviflorum* (E.Salisb.) Farw.
Cypripedium hirsutum var. *parviflorum* (E.Salisb.) Rolfe.
Cypripedium luteum Aiton ex Raf.
Cypripedium luteum var. *parviflorum* (E.Salisb.) Raf.
Cypripedium parviflorum var. *makasin* Farw.
Cypripedium pubescens Willd.
Cypripedium pubescens var. *makasin* Farw.

Distribution: Canada, USA

Cypripedium parviflorum var. **pubescens** (Willd.) Knight

Distribution: Canada, USA

Cypripedium passerinum Richardson
Cypripedium passerinum var. *minganense* Vict.

Distribution: Canada, USA

Cypripedium plectrochilum Franch.
Cypripedium arietinum auct. non. R.Br.

Distribution: China

Cypripedium reginae Walter
Calceolus reginae (Walter) Nieuwl.
Cypripedium album Aiton
Cypripedium calceolus var. *gamma* L.
Cypripedium canadense F.Michx.

Cypripedium reginae var. *album* Rolfe
Cypripedium spectabile E.Salisb.

Distribution: Canada, USA

Cypripedium segawai Masam.
Cypripedium guttatum var. *segawai* (Masam.) Ying

Distribution: China (Taiwan)

Cypripedium shanxiense S.C.Chen

Distribution: China, Japan, Russian Federation

Cypripedium smithii Schltr.
Cypripedium pulchrum Ames & Schltr.

Distribution: China

Cypripedium subtropicum S.C.Chen & K.Y.Lang

Distribution: China

Cypripedium tibeticum King ex Hemsl.
Cypripedium amesianum Williams
Cypripedium compactum Schltr. ex Limpr.
Cypripedium corrugatum Franch.

Distribution: China

Cypripedium wardii Rolfe

Distribution: China

Cypripedium wumengense S.C.Chen

Distribution: China

Cypripedium yatabeanum Makino
Cypripedium guttatum var. *yatabeanum* (Makino) Pfitzer
Cypripedium guttatum subsp. *yatabeanum* (Makino) Hulten

Distribution: Japan, Russian Federation, USA

Cypripedium yunnanense Franch.

Distribution: China

LAELIA BINOMIALS IN CURRENT USAGE

Laelia alaorii F.G.Brieger & Bicalho

Distribution: Brazil

Laelia albida Bateman ex Lindl.
Bletia albida (Bateman ex Lindl.) Rchb.f.
Cattleya albida (Bateman ex Lindl.) Beer
Laelia discolor A.Rich. & Galeotti

Distribution: Mexico

Laelia anceps Lindl.
Amalias anceps (Lindl.) Hoffmanns.
Bletia anceps (Lindl.) Rchb.f.
Cattleya anceps (Lindl.) Beer
Laelia anceps var. *barkeriana* Lindl.
Laelia barkeriana Knowles & Westcroft

Distribution: Mexico

Laelia angereri Pabst

Distribution: Brazil

Laelia autumnalis (La Llave & Lex.) Lindl.
Bletia autumnalis La Llave & Lexarza
Cattleya autumnalis (La Llave & Lexarza) Beer

Distribution: Mexico

Laelia bahiensis Schltr.
Hoffmannseggella bahiensis (Schltr.) H.G.Jones

Distribution: Brazil

Laelia bancalarii R.Gonzalez & Hágsater

Distribution: Mexico

Laelia blumenscheinii Pabst

Distribution: Brazil

Part II: Laelia

Laelia bradei Pabst

Distribution: Brazil

Laelia briegeri Blumensch. ex Pabst

Distribution: Brazil

Laelia cardimii Pabst & A.F.Mello

Distribution: Brazil

Laelia caulescens Lindl.
Bletia caulescens (Lindl.) Rchb.f.
Bletia caulescens var. *libonis* Rchb.f.
Hoffmannseggella caulescens (Lindl.) H.G.Jones

Distribution: Brazil

Laelia cinnabarina Bateman ex Lindl.
Amalias cinnabarina (Bateman ex Lindl.) Hoffmanns.
Bletia cinnabarina (Bateman ex Lindl.) Rchb.f.
Bletia cinnabarina var. *sellowii* Rchb.f.
Cattleya cinnabarina (Bateman ex Lindl.) Beer
Hoffmannseggella cinnabarina (Bateman ex Lindl.) H.G.Jones

Distribution: Brazil

Laelia cowanii Cogn.
Hoffmannseggella brevicaulis H.G.Jones
Laelia brevicaulis (H.G. Jones) Withner

Distribution: Brazil

Laelia crispa (Lindl.) Rchb.f.
Bletia crispa (Lindl.) Rchb.f.
Cattleya crispa Lindl.
Cattleya reflexa Parmentier ex Rchb.f.

Distribution: Brazil

Laelia crispata (Thunb.) Garay
Bletia rupestris (Lindl.) Rchb.f.
Cymbidium crispatum Thunb.
Hoffmannseggella crispata (Thunb.) H.G.Jones
Laelia rupestris Lindl.

Distribution: Brazil

Laelia crispilabia (A.Rich. ex Rchb.f.) Warner
Bletia crispilabia Rchb.f.
Hoffmannseggella crispilabia (Rchb.f.) H.G.Jones
Laelia cinnabarina var. *crispilabia* (A.Rich) Veitch
Laelia lawrenceana hort. ex R.Warner

Distribution: Brazil

Laelia dayana Rchb.f.
Laelia pumila var. *dayana* Burbridge ex Dean

Distribution: Brazil

Laelia duveenii Fowlie

Distribution: Brazil

Laelia elegans (C.Morren) Rchb.f.
Bletia elegans (C.Morren) Rchb.f.
Catlaelia elegans (Rchb.f.) G.Hansen
Cattleya elegans C.Morren
Laelia brysiana Lem.
Laelia devoniensis hort.
Laelia gigantea hort. ex R.Warner
Laelia pachystele Rchb.f.
Laelia purpurata var. *brysiana* Du Buyss
Laelia turneri hort. ex R.Warner
Laeliocattleya elegans (Morren) Rolfe
Laeliocattleya lindenii hort.
Laeliocattleya pachystele (Rchb.f.) Rolfe ex B.S.Williams
Laeliocattleya sayana L.Linden

Distribution: Brazil

Laelia endsfeldzii Pabst

Distribution: Brazil

Laelia esalqueana Blumensch. ex Pabst

Distribution: Brazil

Laelia fidelensis Pabst

Distribution: Brazil

Laelia flava Lindl.
Bletia flava (Lindl.) Rchb.f.
Cattleya flava (Lindl.) Beer

Hoffmannseggella flava (Lindl.) H.G.Jones
Laelia flava var. *aurantiaca* hort.
Laelia fulva Lindl. ex Heynh.
Laelia geraensis Barb.Rodr.

Distribution: Brazil

Laelia furfuracea Lindl.
Bletia furfuracea (Lindl.) Rchb.f.
Cattleya furfuracea (Lindl.) Beer

Distribution: Mexico

Laelia gardneri Pabst ex Zappi

Distribution: Brazil

Laelia ghillanyi Pabst
Hoffmannseggella ghillanyi (Pabst) H.G.Jones

Distribution: Brazil

Laelia gloedeniana Hoehne ex Ruschi
Hoffmannseggella macrobulbosa (Pabst) H.G.Jones
Laelia macrobulbosa Pabst

Distribution: Brazil

Laelia gouldiana Rchb.f.

Distribution: Mexico

Laelia gracilis Pabst

Distribution: Brazil

Laelia grandis Lindl. & Paxton
Bletia grandis (Lindl. & Paxton) Rchb.f.

Distribution: Brazil

Laelia harpophylla Rchb.f.
Bletia harpophylla (Rchb.f.) Rchb.f.
Hoffmannseggella harpophylla (Rchb.f.) H.G.Jones

Distribution: Brazil

Laelia hispidula Pabst & A.F.Mello

Distribution: Brazil

Laelia itambana Pabst

Distribution: Brazil

Laelia jongheana Rchb.f.
Bletia jongheana Rchb.f.

Distribution: Brazil

Laelia kautskyi Pabst
Hoffmannseggella kautskyi (Pabst) H.G.Jones
Laelia harpophylla var. *dulcotensis* hort.
Laelia kautskyana Pabst

Distribution: Brazil

Laelia kettieana Pabst

Distribution: Brazil

Laelia liliputana Pabst
Hoffmannseggella liliputana (Pabst) H.G.Jones

Distribution: Brazil

Laelia lobata (Lindl.) Veitch
Bletia boothiana (Rchb.f.) Rchb.f.
Bletia lobata (Lindl.) Rchb.f.
Cattleya lobata Lindl.
Laelia boothiana Rchb.f.
Laelia grandis var. *purpurea* Rchb.f.
Laelia rivieri Carriere

Distribution: Brazil

Laelia longipes (Rchb.f.) Cogn.
Bletia longipes Rchb.f.

Distribution: Brazil

Laelia lucasiana Rolfe
Bletia lucasiana (Rolfe) Rchb.f.
Laelia longipes var. *alba* Rolfe
Laelia longipes var. *fournieri* (Rchb.f.) Cogn.
Laelia ostermayerii Hoehne ex Ruschi

Part II: Laelia

Distribution: Brazil

Laelia lundii Rchb.f. ex Withner
Bletia lundii Rchb.f. & Warm.
Laelia regnellii Barb.Rodr.
Laelia reichenbachiana H.Wendl. & Kraenzl.

Distribution: Brazil

Laelia mantiqueirae Pabst ex Zappi

Distribution: Brazil

Laelia milleri Blumensch. ex Pabst

Distribution: Brazil

Laelia mixta Hoehne

Distribution: Brazil

Laelia perrinii (Lindl.) Bateman
Bletia perrinii (Lindl.) Rchb.f.
Cattleya integerrima var. *angustifolia* Hook.
Cattleya intermedia var. *angustifolia* Hook.
Cattleya perrinii Lindl.

Distribution: Brazil

Laelia pfisteri Pabst & Senghas

Distribution: Brazil

Laelia pumila (Hook.) Rchb.f.
Bletia pumila (Hook.) Rchb.f.
Cattleya marginata Paxton
Cattleya pinellii hort. ex Lindl.
Cattleya pinellii var. *marginata* Beer
Cattleya pumila Hook.
Laelia praestans var. *nobilis* Lindl.

Distribution: Brazil

Laelia purpurata Lindl.
Bletia purpurata (Lindl.) Rchb.f.
Cattleya brysiana Lem.
Cattleya crispa var. *purpurata* hort. ex Rchb.f.
Cattleya purpurata (Lindl. & Paxton) Beer

Laelia casperiana Rchb.f.
Laelia wyattiana Rchb.f.

Distribution: Brazil

Laelia reginae Pabst

Distribution: Brazil

Laelia rubescens Lindl.
Bletia acuminata (Lindl.) Rchb.f.
Bletia peduncularis (Lindl.) Rchb.f.
Bletia rubescens (Lindl.) Rchb.f.
Bletia violacea (Rchb.f.) Rchb.f.
Cattleya acuminata (Lindl.) Beer
Cattleya peduncularis (Lindl.) Beer
Cattleya rubescens (Lindl.) Beer
Laelia acuminata Lindl.
Laelia peduncularis Lindl.
Laelia pubescens Lem.
Laelia violacea Rchb.f.

Distribution: Costa Rica, El Salvador, Guatemala, Honduras, Mexico, Nicaragua, Panama

Laelia sanguiloba Withner

Distribution: Brazil

Laelia sincorana Schltr.
Cattleya grosvenori Ruschi

Distribution: Brazil

Laelia speciosa (Kunth) Schltr.
Bletia grandiflora La Llave & Lexarza
Bletia speciosa Kunth
Cattleya grahamii Lindl.
Cattleya majalis (Lindl.) Beer
Laelia grandiflora (La Llave & Lexarza) Lindl.
Laelia majalis Lindl.

Distribution: Mexico

Laelia spectabilis (Paxton) Withner
Bletia praestans (Rchb.f.) Rchb.f.
Cattleya pumila var. *major* Lem.
Cattleya spectabilis Paxton
Laelia praestans Rchb.f.
Laelia pumila var. *mirabilis* E.Morren
Laelia pumila var. *praestans* (Rchb.f.) Veitch

Laelia pumila subsp. *praestans* (Rchb.f.) Bicalho

Distribution: Brazil

Laelia tenebrosa (Gower) Rolfe
Laelia grandis var. *tenebrosa* Gower

Distribution: Brazil

Laelia tereticaulis Hoehne ex Ruschi
Hoffmannseggella tereticaulis (Hoehne) H.G.Jones

Distribution: Brazil

Laelia virens Lindl.
Laelia goebeliana Kuepper & Kraenzl.
Laelia johniana Schltr.

Distribution: Brazil

Laelia xanthina Lindl. ex Hook.
Bletia flabellata Rchb.f.
Bletia xanthina (Lindl. ex Hook.) Rchb.f.
Laelia wetmorei Ruschi

Distribution: Brazil

PAPHIOPEDILUM BINOMIALS IN CURRENT USAGE

Paphiopedilum acmodontum Schoser ex M.W.Wood

Distribution: Philippines

Paphiopedilum adductum Asher
Paphiopedilum elliottianum sensu Fowlie

Distribution: Philippines

Paphiopedilum appletonianum (Gower) Rolfe
Cordula appletoniana (Gower) Rolfe
Cypripedium appletonianum Gower
Cypripedium bullenianum var. *appletonianum* (Gower) Rolfe
Cypripedium poyntzianum O'Brien
Cypripedium waltersianum Kraenzl.
Cypripedium wolterianum Kraenzl.
Paphiopedilum appletonianum var. *poyntziamum* (O'Brien) Pfitzer
Paphiopedilum hainanense Fowlie
Paphiopedilum wolterianum (Kraenzl.) Pfitzer

Distribution: China, Cambodia, Lao People's Democratic Republic, Thailand, Viet Nam

Paphiopedilum argus (Rchb.f.) Stein
Cordula argus (Rchb.f.) Rolfe
Cypripedium argus Rchb.f.
Cypripedium pitcherianum Manda
Paphiopedilum argus var. *sriwaniae* (Koop.) Gruss
Paphiopedilum barbatum var. *argus* hort.
Paphiopedilum sriwaniae Koop.

Distribution: Philippines

Paphiopedilum armeniacum S.C.Chen & F.Y.Liu

Distribution: China

Paphiopedilum barbatum (Lindl.) Pfitzer
Cordula barbata (Lindl.) Rolfe
Cordula nigrita (Rchb.f.) Merr.
Cypripedium barbatum Lindl.
Cypripedium barbatum var. *biflorum* (Williams) Williams
Cypripedium biflorum Williams
Cypripedium nigritum Rchb.f.
Paphiopedilum barbatum var. *nigritum* (Rchb.f.) Pfitzer
Paphiopedilum nigritum (Rchb.f.) Pfitzer

Distribution: Thailand, Malaysia

Part II: Paphiopedilum

Paphiopedilum barbigerum T.Tang & F.T.Wang
Paphiopedilum insigne var. *barbigerum* (T.Tang & F.T.Wang) Braem

Distribution: China

Paphiopedilum bellatulum (Rchb.f.) Stein
Cordula bellatula (Rchb.f.) Rolfe
Cypripedium bellatulum Rchb.f.

Distribution: China, Lao People's Democratic Republic, Myanmar, Thailand

Paphiopedilum bougainvilleanum Fowlie

Distribution: Papua New Guinea (Bougainville)

Paphiopedilum bullenianum (Rchb.f.) Pfitzer
Cordula amabilis (Hallier) Merr.
Cordula bulleniana (Rchb.f.) Rolfe
Cypripedium bullenianum Rchb.f.
Cypripedium hookerae var. *amabile* (Hallier) Kraenzl.
Cypripedium hookerae var. *bullenianum* (Rchb.f.) Veitch
Cypripedium robinsonii Ridl.
Paphiopedilum amabile Hallier
Paphiopedilum hookerae var. *bullenianum* (Rchb.f.) Kerch.
Paphiopedilum johorense Fowlie & Yap
Paphiopedilum linii Schoser
Paphiopedilum robinsonii (Ridl.) Ridl.
Paphiopedilum tortipetalum Fowlie

Distribution: Indonesia, Malaysia

Paphiopedilum bullenianum var. **celebesense** (Fowlie & Birk) P.J.Cribb
Paphiopedilum ambonensis hort.
Paphiopedilum celebesense Fowlie & Birk
Paphiopedilum ceramensis Birk

Distribution: Indonesia

Paphiopedilum callosum (Rchb.f.) Stein
Cordula callosa (Rchb.f.) Rolfe
Cypripedium barbatum var. *crossii* hort. ex Veitch
Cypripedium callosum Rchb.f.
Cypripedium crossii Morren
Cypripedium schmidtianum Kraenzl.
Paphiopedilum callosum var. *angustipetalum* Guill.
Paphiopedilum callosum var. *schmidtianum* Kraenzl.
Paphiopedilum orbum Rchb.f.
Paphiopedilum reflexum hort. ex Stein
Paphiopedilum regnieri hort. ex Stein

Distribution: Cambodia, Lao People's Democratic Republic, Thailand, Viet Nam

Paphiopedilum callosum var. **sublaeve** (Rchb.f.) P.J.Cribb
Cypripedium barbatum var. *warneri* hort.
Cypripedium barbatum var. *warnerianum* T.Moore
Cypripedium callosum var. *sublaeve* Rchb.f.
Paphiopedilum birkii Birk
Paphiopedilum callosum subsp. *sublaeve* (Rchb.f.) Fowlie
Paphiopedilum sublaeve (Rchb.f.) Fowlie
Paphiopedilum thailandense Fowlie

Distribution: Thailand, Malaysia

Paphiopedilum charlesworthii (Rolfe) Pfitzer
Cordula charlesworthii (Rolfe) Rolfe
Cypripedium charlesworthii Rolfe

Distribution: Myanmar, Thailand

Paphiopedilum ciliolare (Rchb.f.) Stein
Cordula ciliolaris (Rchb.f.) Rolfe
Cypripedium ciliolare Rchb.f.
Cypripedium ciliolare var. *miteauanum* Linden
Cypripedium miteauanum Linden
Paphiopedilum ciliolare var. *miteauanum* (Linden) Pfitzer
Paphiopedilum superbiens subsp. *ciliolare* (Rchb.f.) M.W.Wood

Distribution: Philippines

Paphiopedilum concolor (Lindl.) Pfitzer
Cordula concolor (Lindl.) Rolfe
Cypripedium concolor Lindl.
Cypripedium tonkinense Godefroy

Distribution: China, Cambodia, Lao People's Democratic Republic, Myanmar, Thailand, Viet Nam

Paphiopedilum dayanum (Lindl.) Stein
Cordula dayana (Lindl.) Rolfe
Cordula petri (Rchb.f.) Rolfe
Cypripedium burbidgei Rchb.f.
Cypripedium dayanum (Lindl.) Rchb.f.
Cypripedium dayi Stone
Cypripedium ernestianum hort.
Cypripedium peteri D.Dean
Cypripedium petri Rchb.f.
Cypripedium spectabile var. *dayanum* Lindl.
Cypripedium superbiens var. *dayanum* (Lindl.) Rchb.f.
Cypripedium × *petri* (Rchb.f.) Rolfe
Cypripedium × *petri* var. *burbidgei* (Rchb.f.) Rolfe
Paphiopedilum burbidgei (Rchb.f.) Pfitzer
Paphiopedilum dayanum var. *petri* (Rchb.f.) Pfitzer
Paphiopedilum petri (Rchb.f.) Rolfe

Distribution: Malaysia

Paphiopedilum delenatii Guillaumin
Cypripedium delenatii (Guillaumin) C.H. Curtis

Distribution: Viet Nam

Paphiopedilum dianthum T.Tang & F.T.Wang
Paphiopedilum parishii var. *dianthum* (T.Wang & F.T.Wang) P.J.Cribb & C.Z.Tang

Distribution: China

Paphiopedilum druryi (Bedd.) Stein
Cordula druryi (Bedd.) Rolfe
Cypripedium druryi Bedd.

Distribution: India

Paphiopedilum emersonii Koop. & P.J.Cribb

Distribution: China

Paphiopedilum exul (Ridl.) Rolfe
Cypripedium exul (Ridl.) Rolfe
Cypripedium insigne var. *exul* Ridl.
Paphiopedilum insigne var. *exul* (Ridley) Braem

Distribution: Thailand

Paphiopedilum fairrieanum (Lindl.) Stein
Cordula fairrieana (Lindl.) Rolfe
Cypripedium fairrieanum Lindl.

Distribution: Bhutan, India

Paphiopedilum fowliei Birk
Paphiopedilum hennisianum var. *fowliei* (Birk) P.J.Cribb

Distribution: Philippines

Paphiopedilum glanduliferum (Blume) Stein
Cordula glandulifera (Blume) Rolfe
Cypripedium gardineri Guillemard
Cypripedium glanduliferum Blume
Cypripedium praestans Rchb.f.
Cypripedium praestans var. *kimballianum* Linden & Rodigas
Paphiopedilum gardineri (Guillemard) Pfitzer
Paphiopedilum glanduliferum var. *gardineri* (Guillemard) Braem
Paphiopedilum glanduliferum var. *praestans* (Rchb.f.) Braem

Paphiopedilum praestans (Rchb.f.) Pfitzer
Paphiopedilum praestans var. *kimballianum* (Linden & Rodigas) Pfitzer

Distribution: Indonesia, Papua New Guinea

Paphiopedilum glanduliferum var. **wilhelminae** (L.O.Williams) P.J.Cribb
Paphiopedilum bodegomii Fowlie
Paphiopedilum gardineri sensu Kennedy non Guillemard
Paphiopedilum praestans subsp. *wilhelminae* (L.O.Williams) M.W.Wood
Paphiopedilum wilhelminae L.O.Williams

Distribution: Indonesia, Papua New Guinea

Paphiopedilum glaucophyllum J.J.Sm.
Cordula glaucophylla (J.J.Sm.) Rolfe
Cypripedium glaucophyllum (J.J.Sm.) Masters
Paphiopedilum victoria-reginae subsp. *glaucophyllum* (J.J.Sm.) M.W.Wood

Distribution: Indonesia

Paphiopedilum glaucophyllum var. **moquetteanum** J.J.Sm.
Paphiopedilum moquetteanum (J.J.Sm.) Fowlie
Paphiopedilum victoria-regina subsp. *glaucophyllum* (J.J.Sm.) M.Wood

Distribution: Indonesia

Paphiopedilum godefroyae (God.-Leb.) Stein
Cordula godefroyae (God.-Leb.) Rolfe
Cypripedium concolor var. *godefroyae* (God.-Leb.) Collett & Hemsl.
Cypripedium godefroyae God.-Leb.
Paphiopedilum × *godefroyae* (God.-Leb.) Braem

Distribution: Thailand

Paphiopedilum godefroyae var. **leucochilum** (Masters) Hallier
Cypripedium godefroyae var. *leucochilum* Rolfe
Paphiopedilum leucochilum (Rolfe) Fowlie
Paphiopedilum × *godefroyae* var. *leucochilum* (Masters) Braem

Distribution: Thailand

Paphiopedilum gratrixianum (Masters) Guillaumin
Cypripedium gratrixianum Masters
Paphiopedilum affine De Wild.
Paphiopedilum villosum var. *affine* (De Wild.) Braem
Paphiopedilum villosum var. *gratrixianum* (Masters) Braem

Distribution: Lao People's Democratic Republic, Viet Nam

Part II: Paphiopedilum

Paphiopedilum haynaldianum (Rchb.f.) Stein
Cordula haynaldiana (Rchb.f.) Rolfe
Cypripedium haynaldianum Rchb.f.

Distribution: Philippines

Paphiopedilum hennisianum (M.W.Wood) Fowlie

Distribution: Philippines

Paphiopedilum henryanum Braem
Paphiopedilum dollii Lueckel

Distribution: China, Viet Nam

Paphiopedilum hirsutissimum (Lindl. ex Hook.) Stein
Cordula hirsutissimum (Lindl. ex Hook.) Rolfe
Cypripedium hirsutissimum Lindl. ex Hook.
Paphiopedilum chiwuanum T.Tang & F.T.Wang
Paphiopedilum hirsutissimum var. *chiwuanum* (T.Tang & F.T.Wang) P.J.Cribb

Distribution: China, India, Myanmar

Paphiopedilum hirsutissimum var. **esquirolei** (Schltr.) P.J.Cribb

Distribution: China, Thailand, Viet Nam

Paphiopedilum hookerae (Rchb.f.) Stein
Cordula hookerae (Rchb.f.) Rolfe
Cypripedium hookerae Rchb.f.

Distribution: Indonesia, Malaysia

Paphiopedilum hookerae var. **volonteanum** (Sander ex Rolfe) Kerch.
Cypripedium hookerae var. *volonteanum* Sander ex Rolfe
Cypripedium volonteanum Sander
Paphiopedilum volonteanum (Sander ex Rolfe) Pfitzer

Distribution: Malaysia

Paphiopedilum insigne (W.Wall ex Lindl.) Pfitzer
Cordula insignis (Lindl.) Raf.
Cypripedium insigne W.Wall ex Lindl.

Distribution: India, Nepal

Paphiopedilum javanicum (Reinw. ex Lindl.) Pfitzer
Cordula javanica (Reinw. ex Lindl.) Rolfe

Cypripedium javanicum Reinw. ex Lindl.

Distribution: Indonesia, Malaysia

Paphiopedilum javanicum var. **virens** (Rchb.f.) Stein
Cypripedium javanicum var. *virens* (Rchb.f.) Veitch
Cypripedium virens Rchb.f.
Paphiopedilum purpurascens Fowlie
Paphiopedilum virens (Rchb.f.) Pfitzer

Distribution: Malaysia

Paphiopedilum kolopakingii Fowlie
Paphiopedilum topperi Braem & H.Mohr

Distribution: Indonesia

Paphiopedilum lawrenceanum (Rchb.f.) Pfitzer
Cordula lawrenceana (Rchb.f.) Merr.
Cypripedium lawrenceanum Rchb.f.
Paphiopedilum barbatum subsp. *lawrenceanum* (Rchb.f.) M.W.Wood

Distribution: Malaysia

Paphiopedilum liemianum (Fowlie) Karas. & K.Saito
Paphiopedilum chamberlainianum subsp. *liemianum* Fowlie
Paphiopedilum chamberlainianum var. *liemianum* (Fowlie) Braem
Paphiopedilum victoria-reginae subsp. *liemianum* (Fowlie) M.W.Wood

Distribution: Indonesia

Paphiopedilum lowii (Lindl.) Stein
Cordula lowii (Lindl.) Rolfe
Cypripedium cruciforme Zoll & Morren
Cypripedium lowii Lindl.

Distribution: Indonesia, Malaysia

Paphiopedilum lowii var. **richardianum** (Asher & Beaman) Gruss
Paphiopedilum richardianum (Asher & Beaman)

Distribution: Indonesia

Paphiopedilum malipoense Chen & Z.H.Tsi

Distribution: China, Viet Nam

Paphiopedilum mastersianum (Rchb.f.) Stein
Cordula mastersiana (Rchb.f.) Rolfe

Cypripedium mastersianum Rchb.f.

Distribution: Indonesia

Paphiopedilum micranthum T.Tang & C.W.Wang

Distribution: China, Viet Nam

Paphiopedilum mohrianum Braem

Distribution: Indonesia

Paphiopedilum niveum (Rchb.f.) Stein
Cordula nivea (Rchb.f.) Rolfe
Cypripedium niveum Rchb.f.
Paphiopedilum concolor var. *niveum* Rchb.f.

Distribution: Thailand, Malaysia

Paphiopedilum papuanum (Ridl.) Ridl.
Cypripedium papuanum Ridl.
Paphiopedilum zieckianum Schoser

Distribution: Indonesia, Papua New Guinea

Paphiopedilum parishii (Rchb.f.) Stein
Cordula parishii (Rchb.f.) Rolfe
Cypripedium parishii Rchb.f.
Selenipedium parishii Andre

Distribution: China, Myanmar, Thailand

Paphiopedilum philippinense (Rchb.f.) Stein
Cordula philippinensis (Rchb.f.) Rolfe
Cypripedium cannartianum Linden
Cypripedium laevigatum Bateman
Cypripedium philippinense Rchb.f.
Cypripedium roebelinii var. *cannartianum* Linden
Paphiopedilum laevigatum (Bateman) Pfitzer
Paphiopedilum philippinense var. *cannartianum* (Linden) Pfitzer
Selenipedium laevigatum (Bateman) May

Distribution: Malaysia, Philippines

Paphiopedilum philippinense var. **roebelenii** (Veitch) P.J.Cribb
Cypripedium philippinense var. *roebelenii* Veitch
Cypripedium roebelenii Rchb.f.
Paphiopedilum roebelenii (Rchb.f.) Pfitzer

Distribution: Philippines

Paphiopedilum primulinum M.W.Wood & Taylor
Paphiopedilum chamberlainianum var. *primulinum* (M.W.Wood & P. Taylor) Braem
Paphiopedilum liemianum var. *primulinum* (M.W.Wood & Taylor) Karas. & K.Saito
Paphiopedilum victoria-reginae var. *primulinum* (M.W.Wood & Taylor) M.W.Wood

Distribution: Indonesia

Paphiopedilum primulinum var. **purpurascens** (M.W.Wood) P.J.Cribb

Distribution: Indonesia

Paphiopedilum purpuratum (Lindl.) Stein
Cordula purpurata (Lindl.) Rolfe
Cypripedium purpuratum Lindl.
Cypripedium sinicum Hance ex Rchb.f.
Paphiopedilum sinicum (Hance ex Rchb.f.) Stein

Distribution: China, Hong Kong, Viet Nam

Paphiopedilum randsii Fowlie

Distribution: Philippines

Paphiopedilum rothschildianum (Rchb.f.) Stein
Cordula rothschildiana (Rchb.f.) Merr.
Cypripedium elliottianum O'Brien
Cypripedium neo-guineense Linden
Cypripedium rothschildianum Rchb.f.
Paphiopedilum elliottianum (O'Brien) Stein
Paphiopedilum nicholsonianum ex hort.
Paphiopedilum rothschildianum var. *elliottianum* (O'Brien) Pfitzer
Selenipedium elliottianum Gower

Distribution: Malaysia

Paphiopedilum sanderianum (Rchb.f.) Stein
Cordula sanderiana Rolfe
Cypripedium sanderianum Rchb.f.

Distribution: Malaysia

Paphiopedilum sangii Braem

Distribution: Indonesia

Paphiopedilum schoseri Braem & H.Mohr
Paphiopedilum bacanum Schoser & Deelder

Distribution: Indonesia, Malaysia

Paphiopedilum spicerianum (Rchb.f. ex Masters & T.Moore) Pfitzer
Cordula spiceriana (Rchb.f. ex Masters & T.Moore) Rolfe
Cypripedium spicerianum Rchb.f. ex Masters & T.Moore

Distribution: India, Myanmar

Paphiopedilum stonei (Hook.) Stein
Cordula stonei (Hook.) Merr.
Cypripedium stonei Hook.

Distribution: Malaysia

Paphiopedilum sukhakulii Schoser & Senghas

Distribution: Thailand

Paphiopedilum supardii Braem & Loeb
Paphiopedilum devogelii Schoser & Deelder
Paphiopedilum 'victoria' De Vogel

Distribution: Indonesia

Paphiopedilum superbiens (Rchb.f.) Stein
Cordula superbiens (Rchb.f.) Rolfe
Cypripedium barbatum var. *superbum* Morren
Cypripedium barbatum var. *veitchii* Lem.
Cypripedium superbiens Rchb.f.
Cypripedium veitchianum hort. ex Lem.

Distribution: Indonesia

Paphiopedilum superbiens var. **curtisii** (Rchb.f.) G.J.Braem
Cordula curtisii (Rchb.f.) Rolfe
Cypripedium curtisii Rchb.f.
Paphiopedilum curtisii (Rchb.f.) Stein

Distribution: Indonesia

Paphiopedilum tigrinum Koop. & N.Haseg.
Paphiopedilum markianum Fowlie

Distribution: China

Paphiopedilum tonsum (Rchb.f.) Stein
Cordula tonsa (Rchb.f.) Rolfe
Cypripedium tonsum Rchb.f.

Distribution: Indonesia

Paphiopedilum tonsum var. **braemii** (Mohr) Gruss
Paphiopedilum braemii Mohr

Distribution: Indonesia

Paphiopedilum urbanianum Fowlie

Distribution: Philippines

Paphiopedilum venustum (Wall. ex Sims) Pfitzer ex Stein
Cordula venusta (Wall.) Rolfe
Cypripedium pardinum (Rchb.f.)
Cypripedium venustum Wall.
Cypripedium venustum Wall. ex Sims
Paphiopedilum pardinum (Rchb.f.) Pfitzer
Paphiopedilum venustum var. *pardinum* (Rchb.f.) Pfitzer
Stimegas venustum (Wall. ex Sims) Raf.

Distribution: Bhutan, India, Nepal

Paphiopedilum victoria-mariae (Sander ex Mast.) Rolfe
Cordula victoria-mariae (Sander ex Mast.) Rolfe
Cypripedium victoria-mariae Sander ex Mast.

Distribution: Indonesia

Paphiopedilum victoria-regina (Sander) M.W.Wood
Cypripedium chamberlainianum Sander
Cypripedium victoria-regina Sander
Paphiopedilum chamberlainianum (Sander) Stein

Distribution: Indonesia

Paphiopedilum villosum (Lindl.) Stein
Cordula villosa (Lindl.) Rolfe
Cypripedium villosum Lindl.

Distribution: India, Myanmar, Thailand

Paphiopedilum villosum var. **annamense** Rolfe

Distribution: Lao People's Democratic Republic, Viet Nam

Paphiopedilum villosum var. **boxallii** (Rchb.f.) Pfitzer
Cordula boxalii (Rchb.f.) Rolfe
Cypripedium boxallii Rchb.f.

Cypripedium boxallii var. *atratum* Masters
Cypripedium dilectum Rchb.f.
Cypripedium villosum var. *boxallii* (Rchb.f.) Veitch
Paphiopedilum boxallii (Rchb.f.) Pfitzer
Paphiopedilum dilectum (Rchb.f.) Pfitzer

Distribution: Myanmar

Paphiopedilum violascens Schltr.
Cordula violascens (Schltr.) Rolfe
Paphiopedilum violascens var. *gautierense* J.J.Sm.

Distribution: Indonesia, Papua New Guinea

Paphiopedilum wardii Summerh.
Cypripedium wardii (Summerh.) C.H.Curtis
Paphiopedilum × *wardii* (Summerh.) Braem

Distribution: Myanmar

Paphiopedilum wentworthianum Schoser & Fowlie
Paphiopedilum dennisii Schoser

Distribution: Papua New Guinea (Bougainville), Solomon Islands

PARAPHALAENOPSIS BINOMIALS IN CURRENT USAGE

Paraphalaenopsis denevei (J.J.Sm.) A.D.Hawkes
Phalaenopsis denevei J.J.Sm.

Distribution: Indonesia

Paraphalaenopsis labukensis Shim, Lamb & Chan
Phalaenopsis labukensis Shim, Lamb & Chan

Distribution: Malaysia

Paraphalaenopsis laycockii (M.R.Hend.) A.D.Hawkes
Phalaenopsis laycockii M.R.Hend.

Distribution: Indonesia

Paraphalaenopsis serpentilingua (J.J.Sm.) A.D.Hawkes
Phalaenopsis denevei var. *alba* Price
Phalaenopsis serpentilingua J.J.Sm.
Phalaenopsis simonsei Simonse

Distribution: Indonesia, Malaysia

Paraphalaenopsis × thorntonii (Holttum) A.D.Hawkes
Phalaenopsis × thorntonii Holttum

Distribution: Indonesia, Malaysia

PHALAENOPSIS BINOMIALS IN CURRENT USAGE

Phalaenopsis amabilis (L.) Blume
Angraecum album-majus Rumph.
Cymbidium amabile (L.) Roxb.
Epidendrum amabile (L.) Roxb.
Phalaenopsis amabilis var. *'gloriosa'* (Pleihari) Van Brero
Phalaenopsis amabilis var. *aurea* (hort.) Rolfe
Phalaenopsis amabilis var. *fournieri* Cogn.
Phalaenopsis amabilis var. *gracillima* Burb.
Phalaenopsis amabilis var. *grandiflora* (Lindl.) Bateman
Phalaenopsis amabilis var. *ramosa* Van Deventer
Phalaenopsis amabilis var. *rimestadiana* Linden
Phalaenopsis amabilis var. *rimestadiana-alba* hort.
Phalaenopsis aphrodite var. *gloriosa* (Rchb.f.) J.Veitch
Phalaenopsis gloriosa Rchb.f.
Phalaenopsis grandiflora Lindl.
Phalaenopsis pleihary Burgeff
Phalaenopsis rimestadiana (Linden) Rolfe
Synadena amabilis (L.) Raf.

Distribution: Brunei Darussalam, Indonesia, Malaysia, Philippines

Phalaenopsis amabilis var. **moluccana** Schltr.
Phalaenopsis amabilis var. *cinerascens* J.J.Sm.
Phalaenopsis amabilis var. *ruckeri* J.J.Sm.
Phalaenopsis celebica Vloten

Distribution: Indonesia

Phalaenopsis amboinensis J.J.Sm.
Phalaenopsis hombronii Finet
Phalaenopsis psilantha Schltr.
Polychilos amboinensis (J.J.Sm.) Shim

Distribution: Indonesia

Phalaenopsis aphrodite Rchb.f.
Phalaenopsis amabilis Lindl.
Phalaenopsis amabilis var. *ambigua* (Rchb.f.) Burb.
Phalaenopsis amabilis var. *aphrodite* (Rchb.f.) Ames
Phalaenopsis amabilis var. *dayana* hort. ex R.Warner & B.S.Williams.
Phalaenopsis amabilis var. *erubescens* (Burb.) Burb.
Phalaenopsis amabilis var. *formosa* Shimadzu
Phalaenopsis amabilis var. *fuscata* (Rchb.f.) Ames
Phalaenopsis amabilis var. *longifolia* Don
Phalaenopsis amabilis var. *rotundifolia* Don
Phalaenopsis ambigua Rchb.f.
Phalaenopsis aphrodite var. *aphrodite* Rchb.f.
Phalaenopsis aphrodite var. *dayana* (hort. ex R.Warner & B.S.Wms.) J.Veitch
Phalaenopsis babuyana Miwa
Phalaenopsis erubescens Burb.

Phalaenopsis formosana Miwa
Phalaenopsis formosum hort. ex Sanders

Distribution: Philippines, China (Taiwan)

Phalaenopsis appendiculata C.E.Carr

Distribution: Malaysia

Phalaenopsis bastianii Gruss & Roellke
Phalaenopsis deltonii Gruss & Roellke

Distribution: Philippines

Phalaenopsis celebensis Sweet

Distribution: Indonesia

Phalaenopsis cochlearis Holttum
Polychilos cochlearis (Holttum) Shim

Distribution: Malaysia

Phalaenopsis corningiana Rchb.f.
Phalaenopsis cumingiana Rchb.f.
Phalaenopsis sumatrana var. *sanguinea* Rchb.f.
Phalaenopsis sumatrana subsp. *sumatrana* Korth. & Rchb.f.
Polychilos corningiana (Rchb.f.) Shim

Distribution: Brunei Darussalam, Indonesia, Malaysia

Phalaenopsis cornucervi (Breda) Blume & Rchb.f.
Phalaenopsis devriesiana Rchb.f.
Polychilos cornucervi Breda
Polystylus cornucervi Hassk.

Distribution: India, Myanmar, Thailand, Brunei Darussalam, Indonesia, Malaysia, Philippines

Phalaenopsis cornucervi var. **picta** (Hassk.) Sweet
Polystylus cornucervi var. *picta* Hassk.

Distribution: Indonesia

Phalaenopsis equestris (Schauer) Rchb.f.
Phalaenopsis riteiwanensis Masamune
Phalaenopsis rosea Lindl.
Phalaenopsis rosea var. *deliciosa* Burb.
Phalaenopsis stauroglottis hort. ex Rollinson

Stauroglottis equestris Schauer
Stauroglottis riteiwanensis Masamune

Distribution: China (Taiwan), Philippines

Phalaenopsis equestris var. **alba** hort.

Distribution: Philippines

Phalaenopsis equestris var. **leucaspis** Rchb.f.
Phalaenopsis rosea var. *leucaspis* (Rchb.f.) Rolfe

Distribution: Philippines

Phalaenopsis equestris var. **leucotanthe** Rchb.f. ex God.-Leb.
Phalaenopsis rosea var. *aurantiaca* Gower

Distribution: Philippines

Phalaenopsis equestris var. **rosea** Valmayor & Tiu

Distribution: Philippines

Phalaenopsis fasciata Rchb.f.
Polychilos fasciata (Rchb.f.) Shim

Distribution: Philippines

Phalaenopsis fimbriata J.J.Sm.
Phalaenopsis fimbriata var. *tortilis* Gruss & Roellke
Phalaenopsis gigantea var. *decolorata* Braem
Polychilos fimbriata (J.J.Sm.) Shim

Distribution: Indonesia, Malaysia

Phalaenopsis fimbriata var. **sumatrana** J.J.Sm.

Distribution: Indonesia

Phalaenopsis floresensis Fowlie

Distribution: Indonesia

Phalaenopsis fuscata Rchb.f.
Phalaenopsis denisiana Cogn.
Polychilos fuscata (Rchb.f.) Shim

Part II: Phalaenopsis

Distribution: Brunei Darussalam, Indonesia, Malaysia, Philippines

Phalaenopsis × gersenii (Teijsm. & Binn.) Rolfe
Phalaenopsis × singuliflora J.J.Sm.
Phalaenopsis sumatrana var. *gersenii* (Teijsm. & Binn.) Rchb.f.
Phalaenopsis violacea var. *schroederi* hort.
Phalaenopsis violacea var. *schroederi* Rodigas
Phalaenopsis violacea var. *schroederiana* Rchb.f.
Phalaenopsis zebrina var. *gersenii* Teijsm. & Binn.
Phalaenopsis zebrina var. *lilacina* Teijsm. & Binn.

Distribution: Indonesia, Malaysia

Phalaenopsis gibbosa Sweet

Distribution: Lao People's Democratic Republic, Viet Nam

Phalaenopsis gigantea J.J.Sm.
Polychilos gigantea (J.J.Sm.) Shim

Distribution: Brunei Darussalam, Indonesia, Malaysia

Phalaenopsis hieroglyphica (Rchb.f.) Sweet
Phalaenopsis lueddemanniana var. *hieroglyphica* Rchb.f.
Phalaenopsis lueddemanniana var. *palawanensis* Quisumb.
Phalaenopsis lueddemanniana var. *surigaoensis* hort.
Polychilos hieroglyphica (Rchb.f.) Shim

Distribution: Philippines

Phalaenopsis inscriptiosinensis Fowlie

Distribution: Indonesia

Phalaenopsis × intermedia Lindl.
Phalaenopsis × aphroditi-equestri Rchb.f.
Phalaenopsis × brymeriana hort. ex R.Warner & B.S.Williams.
Phalaenopsis delicata Rchb.f.
Phalaenopsis × diezii (Covera) Quisumb.
Phalaenopsis × intermedia var. *diezii* Covera
Phalaenopsis × intermedia var. *porteana* Burb.
Phalaenopsis × intermedia var. *portei* Rchb.f.
Phalaenopsis × intermedia var. *vesta* hort.
Phalaenopsis lobbii hort. ex Rchb.f.
Phalaenopsis × portei (Rchb.f.) Denning
Phalaenopsis × porteri hort. ex Index Kew.
Phalaenopsis × vesta hort.

Distribution: Philippines

Phalaenopsis javanica J.J.Sm.
Phalaenopsis latisepala Rolfe
Polychilos javanica (J.J.Sm.) Shim

Distribution: Indonesia

Phalaenopsis kunstleri Hook.f.
Phalaenopsis fuscata var. *kunstleri* nom. nud.
Polychilos kunstleri (Hook.f.) Shim

Distribution: Myanmar, Malaysia

Phalaenopsis lamelligera Sweet

Distribution: Brunei Darussalam, Indonesia, Malaysia

Phalaenopsis × leucorrhoda Rchb.f.
Phalaenopsis casta W.Boxall ex Naves
Phalaenopsis × *casta* Rchb.f.
Phalaenopsis × *casta* var. *superbissima* hort.
Phalaenopsis × *cynthia* Rolfe
Phalaenopsis × *leucorrhoda* var. *alba* hort.
Phalaenopsis × *leucorrhoda* var. *casta* (Rchb.f.) J.Veitch
Phalaenopsis × *leucorrhoda* var. *cynthia* (Rolfe) J.Veitch
Phalaenopsis × *leucorrhoda* var. *grandiflora* hort.
Phalaenopsis × *rothschildiana* Rchb.f.
Phalaenopsis × *rothschildiana* var. *tatsuta* Iwasaki
Phalaenopsis × *schillerano* Rolfe
Phalaenopsis × *youngiana* hort.
Phalaenopsis × *youngii* hort.

Distribution: Philippines

Phalaenopsis lindenii Loher

Distribution: Philippines

Phalaenopsis lobbii (Rchb.f.) Sweet
Phalaenopsis listeri Berkeley
Phalaenopsis parishii var. *lobbii* Rchb.f.
Polychilos lobbii (Rchb.f.) Shim

Distribution: India, Myanmar, Viet Nam

Phalaenopsis lowii Rchb.f.
Phalaenopsis proboscidioides Parish ex Rchb.f.
Polychilos lowii (Rchb.f.) Shim

Distribution: Myanmar

Part II: Phalaenopsis

Phalaenopsis lueddemanniana Rchb.f.
Phalaenopsis lueddemannii Boxall ex Naves
Polychilos lueddemanniana (Rchb.f.) Shim

Distribution: Philippines

Phalaenopsis lueddemanniana var. **delicata** Rchb.f.

Distribution: Philippines

Phalaenopsis lueddemanniana var. **ochracea** Rchb.f.
Phalaenopsis ochracea (Rchb.f.) Carriere ex Stein

Distribution: Philippines

Phalaenopsis maculata Rchb.f.
Phalaenopsis cruciata Schltr.
Phalaenopsis muscicola Ridl.
Polychilos maculata (Rchb.f.) Shim

Distribution: Brunei Darussalam, Indonesia, Malaysia

Phalaenopsis mannii Rchb.f.
Phalaenopsis boxallii Rchb.f.
Polychilos mannii (Rchb.f.) Shim

Distribution: India, Viet Nam

Phalaenopsis mariae Burb. ex R.Warner & B.S.Williams
Polychilos mariae (Burb. ex R.Warner & B.S.Williams) Shim

Distribution: Indonesia, Malaysia, Philippines

Phalaenopsis micholitzii Rolfe
Polychilos micholitzii (Rolfe) Shim

Distribution: Philippines

Phalaenopsis modesta J.J.Sm.
Polychilos modesta (J.J.Sm.) Shim

Distribution: Indonesia, Malaysia

Phalaenopsis modesta var. **bella** Gruss & Roellke

Distribution: Indonesia, Malaysia

Phalaenopsis mysorensis Saldanha

Distribution: India

Phalaenopsis pallens (Lindl.) Rchb.f.
Phalaenopsis foerstermannii Rchb.f.
Phalaenopsis lueddemanniana var. *pallens* Burb.
Phalaenopsis pallens var. *trullifera* Sweet
Polychilos pallens (Lindl.) Shim
Stauropsis pallens Rchb.f.
Trichoglottis pallens Lindl.

Distribution: Philippines

Phalaenopsis pallens var. **alba** (Ames & Quisumb.) Sweet
Phalaenopsis mariae var. *alba* Ames & Quisumb.

Distribution: Philippines

Phalaenopsis pallens var. **denticulata** (Rchb.f.) Sweet
Phalaenopsis denticulata Rchb.f.

Distribution: Philippines

Phalaenopsis pantherina Rchb.f.
Phalaenopsis luteola Burb.
Polychilos pantherina (Rchb.f.) Shim

Distribution: Indonesia, Malaysia

Phalaenopsis parishii Rchb.f.
Grafia parishii (Rchb.f.) A.D.Hawkes
Polychilos parishii (Rchb.f.) Shim

Distribution: Myanmar, Thailand

Phalaenopsis philippinensis Golamco ex Fowlie & Tang

Distribution: Philippines

Phalaenopsis pulchra (Rchb.f.) Sweet
Phalaenopsis lueddemanniana var. *pulchra* Rchb.f.
Phalaenopsis lueddemanniana subsp. *pulchra* (Rchb.f.) Veitch
Phalaenopsis lueddemanniana var. *purpurea* Ames & Quisumb.
Polychilos pulchra (Rchb.f.) Shim

Distribution: Philippines

Phalaenopsis reichenbachiana Rchb.f. & Sander
Phalaenopsis kimballiana Gower

Distribution: Philippines

Phalaenopsis robinsonii J.J.Sm.

Distribution: Indonesia

Phalaenopsis rosenstromii Bailey
Phalaenopsis amabilis var. *rosenstromii* (Bailey) Nicholls
Phalaenopsis amabilis var. *papuana* Schltr.

Distribution: Australia, Indonesia, Papua New Guinea

Phalaenopsis sanderiana Rchb.f.
Phalaenopsis amabilis var. *sanderiana* (Rchb.f.) J.J.Davis
Phalaenopsis aphrodite var. *sanderae* (Rchb.f.) Quisumb.

Distribution: Philippines

Phalaenopsis sanderiana var. **alba** (J.Veitch) Stein
Phalaenopsis alcicornis Rchb.f.
Phalaenopsis sanderiana subsp. *alba* Veitch
Phalaenopsis sanderiana var. *pulcherrima* Rolfe

Distribution: Philippines

Phalaenopsis sanderiana var. **marmorata** Rchb.f.
Phalaenopsis sanderiana subsp. *marmorata* (Rchb.f.) Veitch
Phalaenopsis sanderiana var. *punctata* O'Brien
Phalaenopsis sanderiana subsp. *punctata* (O'Brien) Veitch

Distribution: Philippines

Phalaenopsis schilleriana Rchb.f.
Phalaenopsis schilleriana var. *compacta-nana* hort.
Phalaenopsis schilleriana var. *delicata* Dean
Phalaenopsis schilleriana var. *odorata* Van Brero
Phalaenopsis schilleriana var. *viridi-maculata* Ducharte

Distribution: Philippines

Phalaenopsis schilleriana var. **immaculata** Rchb.f.
Phalaenopsis curnowiana hort.
Phalaenopsis schilleriana var. *advena* Rchb.f.
Phalaenopsis schilleriana subsp. *immaculata* (Rchb.f.) Veitch

Distribution: Philippines

Phalaenopsis schilleriana var. **purpurea** O'Brien

Distribution: Philippines

Phalaenopsis schilleriana var. **splendens** R.Warner
Phalaenopsis schilleriana var. *grandiflora* Van Brero
Phalaenopsis schilleriana var. *major* Hook.f. ex Rolfe

Distribution: Philippines

Phalaenopsis speciosa Rchb.f.
Phalaenopsis speciosa var. *maculata* Gower
Polychilos speciosa (Rchb.f.) Shim

Distribution: Indonesia

Phalaenopsis speciosa var. **christiana** Rchb.f.
Phalaenopsis speciosa subsp. *christiana* (Rchb.f.) Veitch

Distribution: Indonesia

Phalaenopsis speciosa var. **imperatrix** Rchb.f.
Phalaenopsis imperati Gower
Phalaenopsis speciosa subsp. *imperatrix* (Rchb.f.) Veitch
Phalaenopsis speciosa var. *purpurata* Rchb.f. ex Hook.f.

Distribution: Indonesia

Phalaenopsis stobartiana Rchb.f.
Phalaenopsis wightii var. *stobartiana* (Rchb.f.) Burb.
Phalaenopsis hainanensis Tang & Wang

Distribution: China

Phalaenopsis stuartiana Rchb.f.
Phalaenopsis schilleriana var. *alba* Roebelen
Phalaenopsis schilleriana var. *stuartiana* (Rchb.f.) Burb.
Phalaenopsis schilleriana var. *vestalis* Rchb.f.
Phalaenopsis schilleriana subsp. *vestalis* (Rchb.f.) Veitch
Phalaenopsis stuartiana var. *bella* Rchb.f.
Phalaenopsis stuartiana var. *nobilis* Rchb.f.
Phalaenopsis stuartiana var. *punctatissima* Rchb.f.
Phalaenopsis stuartiana var. *punctulata* Linden

Distribution: Philippines

Phalaenopsis sumatrana Korth. & Rchb.f.
Phalaenopsis acutifolia Linden
Phalaenopsis sumatrana var. *kimballiana* Rchb.f.
Phalaenopsis zebrina Witte

Polychilos sumatrana (Korth. & Rchb.f.) Shim

Distribution: Myanmar, Thailand, Viet Nam, Brunei Darussalam, Indonesia, Malaysia

Phalaenopsis sumatrana var. **alba** G.Wilson

Distribution: Indonesia

Phalaenopsis sumatrana var. **paucivittata** Rchb.f.
Phalaenopsis paucivittata (Rchb.f.) Fowlie

Distribution: Indonesia

Phalaenopsis tetraspis Rchb.f.
Phalaenopsis barrii King ex Hook.f.
Phalaenopsis speciosa var. *tetraspis* (Rchb.f.) Sweet

Distribution: Indonesia, Myanmar

Phalaenopsis thalebanii Seidenfaden

Distribution: Thailand

Phalaenopsis × valentinii Rchb.f.

Distribution: Malaysia

Phalaenopsis × veitchiana Rchb.f.
Phalaenopsis × *gertrudeae* Quisumb.
Phalaenopsis × *intermedia* var. *brymeriana* Rchb.f.
Phalaenopsis × *veitchiana* var. *brachyodon* Rchb.f.
Phalaenopsis veitchii hort.
Phalaenopsis × *virataii* Quisumb.

Distribution: Philippines

Phalaenopsis venosa Shim & Fowlie
Polychilos venosa (Shim & Fowlie) Shim

Distribution: Indonesia, Philippines

Phalaenopsis violacea Witte
Phalaenopsis bellina (Rchb.f.) Christenson
Phalaenopsis violacea var. *bellina* Rchb.f.
Phalaenopsis violacea var. *chloracea* Rchb.f.
Polychilos violacea (Witte) Shim
Staurites violacea Rchb.f.
Stauropsis violacea Rchb.f.

Distribution: Indonesia, Malaysia

Phalaenopsis violacea var. **alba** Teijsm. & Binn.
Phalaenopsis violacea subsp. *alba* (Teijsm. & Binn.) Veitch

Distribution: Indonesia, Malaysia

Phalaenopsis violacea var. **bowringiana** Rchb.f.

Distribution: Indonesia

Phalaenopsis violacea var. **murtoniana** Rchb.f.

Distribution: Malaysia

Phalaenopsis violacea var. **punctata** Rchb.f.

Distribution: Malaysia

Phalaenopsis viridis J.J.Sm.
Phalaenopsis forbesii Ridl.
Polychilos viridis (J.J.Sm.) Shim

Distribution: Indonesia

Phalaenopsis wilsonii Rolfe

Distribution: China

PHRAGMIPEDIUM BINOMIALS IN CURRENT USAGE

Phragmipedium besseae Dodson & E.Kuhn

Distribution: Ecuador, Peru

Phragmipedium boissierianum (Rchb.f.) Rolfe
Cypripedium grandiflorum Pav.
Paphiopedilum boissierianum (Rchb.f.) Stein
Phragmipedium cajamarcae Schltr.
Phragmipedium czerwiakowianum (Rchb.f.) Rolfe
Phragmipedium reticulatum (Rchb.f.) Garay
Selenipedium boissierianum Rchb.f.
Selenipedium duboisii nom. Kewensis
Selenipedium duboissierianum hort.

Distribution: Ecuador, Peru

Phragmipedium caricinum (Lindl. & Paxton) Rolfe
Cypripedium caricinum Lindl. & Paxton
Paphiopedilum caricinum (Lindl. & Paxton) Stein
Selenipedium caricinum (Lindl. & Paxton) Rchb.f.

Distribution: Bolivia

Phragmipedium caudatum (Lindl.) Rolfe
Cypripedium caudatum Lindl.
Cypripedium humboldtii Warsz.
Paphiopedilum caudatum (Lindl.) Pfitzer
Phragmipedium warszewiczianum (Rchb.f.) Garay
Selenipedium caudatum (Lindl.) Rchb.f.

Distribution: Costa Rica, Guatemala, Panama, Bolivia, Ecuador, Peru

Phragmipedium exstaminodium Castano, Hágsater & E.Aguirre

Distribution: Mexico

Phragmipedium hirtzii Dodson

Distribution: Colombia, Ecuador

Phragmipedium klotzschianum (Rchb.f.) Rolfe
Cypripedium klotzschianum Rchb.f. ex M.R.Schomb.
Cypripedium schomburgkianum Klotzsch
Paphiopedilum klotzschianum (Rchb.f.) Stein
Selenipedium klotzschianum Rchb.f.
Selenipedium schomburgkianum (Klotzsch) Desbois

Distribution: Brazil, Guyana, Venezuela

Phragmipedium lindenii (Lindl.) Dressler & N.H.Williams
Cypripedium caudatum var. *lindenii* (Lindl.) Veitch
Cypripedium lindenii (Lindl.) Van Houtte
Paphiopedilum caudatum var. *lindenii* (Lindl.) Stein
Phragmipedium caudatum var. *lindenii* (Lindl.) Pfitzer
Selenipedium caudatum var. *lindenii* (Lindl.) Pucci
Selenipedium caudatum var. *uropedium* (Lindl.) Rolfe
Selenipedium lindenii (Lindl.) Nichols
Uropedium lindenii Lindl.

Distribution: Colombia, Ecuador

Phragmipedium lindleyanum (Lindl.) Rolfe
Cypripedium lindleyanum M.R.Schomb. ex Lindl.
Paphiopedilum lindleyanum (M.R.Schomb.) Stein
Selenipedium lindleyanum (M.R.Schomb. ex Lindl.) Rchb.f.

Distribution: Brazil, French Guiana, Guyana, Suriname, Venezuela

Phragmipedium lindleyanum var. **kaieteurum** (N.E.Br.) Rchb.f. ex Pfitzer
Phragmipedium kaieteurum (N.E.Br.) Garay
Selenipedium kaieteurum N.E.Br.
Selenipedium lindleyanum var. *kaieteurum* (N.E.Br.) Cogn.

Distribution: Venezuela

Phragmipedium longifolium (Rchb.f. & Warsz.) Rolfe
Cypripedium hincksianum Rchb.f.
Cypripedium longifolium Warsz. & Rchb.f.
Paphiopedilum dariense (Rchb.f.) Stein
Paphiopedilum hincksianum (Rchb.f.) Stein
Paphiopedilum longifolium (Rchb.f.) Stein
Phragmipedium dariense (Rchb.f.) Garay
Phragmipedium hartwegii (Rchb.f.) L.O.Williams
Phragmipedium hincksianum (Rchb.f.) Garay
Phragmipedium roezlii (Rchb.f.) Garay
Selenipedium dariense Rchb.f.
Selenipedium longifolium Rchb.f.& Warsz.
Selenipedium reichenbachii Endres

Distribution: Costa Rica, Panama, Colombia, Ecuador

Phragmipedium pearcei (Rchb.f.) Rauh & Senghas
Cypripedium caricinum Bateman
Phragmipedium ecuadorense Garay
Selenipedium pearcei Rchb.f.

Distribution: Ecuador, Peru

Phragmipedium portillae Gruss & Roeth

Distribution: Ecuador

Phragmipedium richteri Roeth & Gruss

Distribution: Peru

Phragmipedium sargentianum (Rolfe) Rolfe
Cypripedium sargentianum (Rolfe) Kraenzl.
Phragmipedilum sargentianum (Rolfe) Rolfe
Phragmopedilum sargentianum (Rolfe) Pfister
Selenipedium sargentianum Rolfe

Distribution: Brazil

Phragmipedium schlimii (Linden & Rchb.f.) Rolfe
Cypripedium schlimii (Rchb.f.) Bateman
Paphiopedilum schlimii (Rchb.f.) Stein
Selenipedium schlimii Linden & Rchb.f.

Distribution: Colombia

Phragmipedium vittatum (Vell.) Rolfe
Cypripedium binoti hort.
Cypripedium paulistanum Barb. Rodr.
Cypripedium vittatum Vell.
Paphiopedilum vittatum (Vell.) Stein
Selenipedium paulistanum (Barb. Rodr.) Rolfe
Selenipedium vittatum (Vell.) Desbois

Distribution: Brazil

Phragmipedium wallisii (Rchb.f.) Garay
Cypripedium caudatum var. *wallisii* (Rchb.f.) Veitch
Paphiopedilum caudatum var. *wallisii* (Rchb.f.) Stein
Selenipedium caudatum var. *wallisii* (Rchb.f.) Pucci
Selenipedium wallisii Rchb.f.

Distribution: Colombia, Ecuador

Phragmipedium xerophyticum J.C.Soto, Salazar & Hágsater
Mexipedium xerophyticum V.Albert & M.W.Chase

Distribution: Mexico

PLEIONE BINOMIALS IN CURRENT USAGE

Pleione albiflora P.J.Cribb & C.Z.Tang
Pleione pinkepankii Braem

Distribution: China, Myanmar

Pleione amoena Schltr.

Distribution: China

Pleione bulbocodioides (Franch.) Rolfe
Coelogyne bulbocodioides Franch.
Coelogyne delavayi Rolfe
Pleione communis Gagnep.
Pleione communis var. *subobtusum* Gagnep.
Pleione delavayi (Rolfe) Rolfe
Pleione fargesii Gagnep.
Pleione ganchuenensis Gagnep.
Pleione mairei Schltr.
Pleione pogonioides (Rolfe) Rolfe
Pleione rhombilabia Hand.-Mazz.
Pleione smithii Schltr.

Distribution: China, China (Tibet), Lao People's Democratic Republic

Pleione chunii C.L.Tso
Pleione aurita P.J.Cribb & Pfennig

Distribution: China

Pleione × confusa P.J.Cribb & C.Z.Tang

Distribution: China

Pleione coronaria P.J.Cribb & C.Z.Tang

Distribution: Nepal

Pleione formosana Hayata
Pleione hui Schltr.
Pleione pricei Rolfe

Distribution: China

Pleione forrestii Schltr.
Pleione alba Li & K.M.Feng

Part II: Pleione

Distribution: China, Myanmar

Pleione grandiflora (Rolfe) Rolfe
Coelogyne grandiflora Rolfe

Distribution: China

Pleione hookeriana (Lindl.) B.S.Williams
Coelogyne hookeriana Lindl.
Coelogyne hookeriana var. *brachyglossa* Rchb.f.
Pleione hookeriana var. *brachyglossa* (Rchb.f.) Rolfe
Pleione laotica Kerr

Distribution: China, China (Tibet), Lao People's Democratic Republic, Thailand, Bhutan, India, Nepal, Myanmar

Pleione humilis (J.E.Sm.) D.Don
Coelogyne humilis (J.E.Sm.) Lindl.
Coelogyne humilis var. *albata* Rchb.f.
Coelogyne humilis var. *tricolor* Rchb.f.
Epidendrum humile Sm.
Pleione diantha Schltr.
Pleione humilis var. *adnata* Pfitzer
Pleione humilis var. *pulchella* E.W.Cooper
Pleione humilis var. *purpurascens* Pfitzer

Distribution: India, Nepal, Myanmar

Pleione × kohlsii Braem

Distribution: China

Pleione × lagenaria Lindl.
Coelogyne lagenaria (Lindl.) Lindl.

Distribution: India

Pleione limprichtii Schltr.

Distribution: China

Pleione maculata (Lindl.) Lindl.
Coelogyne arthuriana Rchb.f.
Coelogyne diphylla (Lindl.) Lindl.
Coelogyne maculata Lindl.
Gymnostylis candida Wall. ex Pfitzer
Pleione diphylla Lindl.
Pleione maculata var. *arthuriana* (Rchb.f.) Rolfe ex Kraenzl.
Pleione maculata var. *virginea* Rchb.f.

Distribution: China, Thailand, Bhutan, India, Myanmar

Pleione praecox (J.E.Sm.) D.Don
Coelogyne birmanica Rchb.f.
Coelogyne praecox (J.E.Sm.) Lindl.
Coelogyne praecox var. *sanguinea* Lindl.
Coelogyne praecox var. *tenera* Rchb.f.
Coelogyne praecox var. *wallichiana* (Lindl.) Lindl.
Coelogyne reichenbachiana T.Moore & Veitch
Coelogyne wallichiana Lindl.
Coelogyne wallichii Hook.
Epidendrum praecox Sm.
Pleione birmanica (Rchb.f.) B.S.Williams
Pleione concolor hort. ex B.S.Williams
Pleione praecox var. *alba* E.Cooper
Pleione praecox var. *birmanica* (Rchb.f.) G.B.Grant
Pleione praecox var. *candida* Pfitzer
Pleione praecox var. *sanguinea* (Lindl.) Pfitzer
Pleione praecox var. *wallichiana* (Lindl.) E.Cooper
Pleione reichenbachiana (T.Moore & Veitch) B.S.Williams
Pleione wallichiana (Lindl.) Lindl.

Distribution: China, Lao People's Democratic Republic, Thailand, Bhutan, India, Nepal, Myanmar

Pleione saxicola T.Tang & Wang ex S.C.Chen

Distribution: China, Bhutan

Pleione scopulorum W.W.Sm.
Bletilla scopulorum (W.W.Sm.) Schltr.

Distribution: China, Myanmar

Pleione speciosa Ames & Schltr.
Coelogyne henryi Rolfe
Coelogyne pogonioides Rolfe
Pleione henryi (Rolfe) Schltr.
Pleione pogonioides (Rolfe) Schltr.

Distribution: China

Pleione yunnanensis (Rolfe) Rolfe
Coelogyne yunnanensis Rolfe
Pleione chiwuana T.Tang & W.Wang

Distribution: China, Myanmar

SOPHRONITELLA BINOMIALS IN CURRENT USAGE

Sophronitella violacea (Lindl.) Schltr.
Cattleya violacea (Lindl.) Beer
Sophronitis violacea Lindl.

Distribution: Brazil

SOPHRONITIS BINOMIALS IN CURRENT USAGE

Sophronitis acuensis Fowlie

Distribution: Brazil

Sophronitis bicolor F.Miranda

Distribution: Brazil

Sophronitis brevipedunculata (Cogn.) Fowlie
Sophronitis wittigiana var. *brevipedunculata* Cogn.

Distribution: Brazil

Sophronitis cernua (Lindl.) Lindl.
Epidendrum humile Vell.
Sophronia cernua (Lindl.) Lindl.
Sophronitis hoffmannseggii Rchb.f.
Sophronitis modesta Lindl.
Sophronitis nutans Hoffmanns.

Distribution: Brazil, Paraguay

Sophronitis coccinea (Lindl.) Rchb.f.
Cattleya coccinea Lindl.
Cattleya grandiflora (Lindl.) Beer
Eunannos coccinea (Lindl.) Porto & Brade
Sophronitis coccinea var. *rossiteriana* (Barb. Rodr.) Pabst & Dungs
Sophronitis grandiflora Lindl.
Sophronitis lowii hort. ex Curtis
Sophronitis militaris Rchb.f.
Sophronitis rossiteriana Barb. Rodr.

Distribution: Brazil

Sophronitis mantiqueirae (Fowlie) Fowlie
Sophronitis coccinea subsp. *mantiqueirae* Fowlie
Sophronitis coccinea var. *parviflora* hort.

Distribution: Brazil

Sophronitis pterocarpa Lindl. & Paxt.
Cattleya pterocarpa (Lindl. & Paxt.) Beer
Sophronitis cernua var. *pterocarpa* hort.

Distribution: Brazil, Paraguay

Part II: Sophronitis

Sophronitis pygmaea (Pabst) Withner
Sophronitis coccinea subsp. *pygmaea* Pabst

Distribution: Brazil

Sophronitis wittigiana Barb. Rodr.
Sophronitis grandiflora var. *rosea* hort.
Sophronitis purpurea Rchb.f
Sophronitis rosea hort. ex Gastling
Sophronitis violacea O'Brien

Distribution: Brazil

PART III: COUNTRY CHECKLIST FOR THE GENERA:

Cattleya, Constantia, Cypripedium, Laelia, Paphiopedilum, Paraphalaenopis, Phalaenopsis, Phragmipedium, Pleione, Sophronitella and *Sophronitis*

PART III: COUNTRY CHECKLIST FOR THE GENERA: *Cattleya, Constantia, Cypripedium, Laelia, Paphiopedilum, Paraphalaenopsis, Phalaenopsis, Phragmipedium, Pleione, Sophronitella* and *Sophronitis*

ARGENTINA

Cattleya loddigesii Lindl.

AUSTRALIA

Phalaenopsis rosenstromii Bailey

AUSTRIA

Cypripedium calceolus L.

BELGIUM

Cypripedium calceolus L.

BELIZE

Cattleya bowringiana O'Brien

BHUTAN

Cypripedium cordigerum D.Don
Cypripedium elegans Rchb.f.
Cypripedium guttatum Sw.
Cypripedium himalaicum Rolfe ex Hemsl.
Paphiopedilum fairrieanum (Lindl.) Stein
Paphiopedilum venustum (Wall. ex Sims) Pfitzer ex Stein
Phalaenopsis lobbii (Rchb.f.) Sweet
Pleione hookeriana (Lindl.) B.S.Williams
Pleione maculata (Lindl.) Lindl.
Pleione praecox (J.E.Sm.) D.Don
Pleione saxicola T.Tang & Wang ex S.C.Chen

BOLIVIA

Cattleya luteola Lindl.
Phragmipedium caricinum (Lindl. & Paxton) Rolfe
Phragmipedium caudatum (Lindl.) Rolfe

BRAZIL

Cattleya aclandiae Lindl.
Cattleya amethystoglossa Linden & Rchb.f. ex R.Warner
Cattleya araguaiensis Pabst
Cattleya bicolor Lindl.
Cattleya dolosa (Rchb.f.) Rchb.f.
Cattleya dormaniana (Rchb.f.) Rchb.f.
Cattleya eldorado Linden ex Van Houtte
Cattleya elongata Barb.Rodr.
Cattleya forbesii Lindl.

BRAZIL (continued)

Cattleya granulosa Lindl.
Cattleya guttata Lindl.
Cattleya harrisoniana Bateman ex Lindl.
Cattleya intermedia Graham ex Hook.
Cattleya kerrii F.G.Brieger & Bicalho
Cattleya labiata Lindl.
Cattleya leopoldii Verschaffelt ex Lem.
Cattleya loddigesii Lindl.
Cattleya luteola Lindl.
Cattleya nobilior Rchb.f.
Cattleya porphyroglossa Linden & Rchb.f.
Cattleya schilleriana Rchb.f.
Cattleya schofeldiana Rchb.f.
Cattleya tenuis Campacci & Vedovello
Cattleya velutina Rchb.f.
Cattleya violacea (Kunth) Rolfe
Cattleya walkeriana Gardner
Cattleya warneri T.Moore ex R.Warner
Constantia australis (Cogn.) Porto & Brade
Constantia cipoensis Porto & Brade
Constantia cristinae F.Miranda
Constantia microscopica F.Miranda
Constantia rupestris Barb. Rodr.
Laelia alaorii F.G.Brieger & Bicalho
Laelia angereri Pabst
Laelia bahiensis Schltr.
Laelia blumenscheinii Pabst
Laelia bradei Pabst
Laelia briegeri Blumensch. ex Pabst
Laelia cardimii Pabst & A.F.Mello
Laelia caulescens Lindl.
Laelia cinnabarina Bateman ex Lindl.
Laelia cowanii Cogn.
Laelia crispa (Lindl.) Rchb.f.
Laelia crispata (Thunb.) Garay
Laelia crispilabia (A.Rich. ex Rchb.f.) Warner
Laelia dayana Rchb.f.
Laelia duveenii Fowlie
Laelia elegans (C.Morren) Rchb.f.
Laelia endsfeldzii Pabst
Laelia esalqueana Blumensch. ex Pabst
Laelia fidelensis Pabst
Laelia flava Lindl.
Laelia gardneri Pabst ex Zappi
Laelia ghillanyi Pabst
Laelia gloedeniana Hoehne ex Ruschi
Laelia gracilis Pabst
Laelia grandis Lindl. & Paxton
Laelia harpophylla Rchb.f.
Laelia hispidula Pabst & A.F.Mello
Laelia itambana Pabst
Laelia jongheana Rchb.f.
Laelia kautskyi Pabst
Laelia kettieana Pabst

BRAZIL (continued)

Laelia liliputana Pabst
Laelia lobata (Lindl.) Veitch
Laelia longipes (Rchb.f.) Cogn.
Laelia lucasiana Rolfe
Laelia lundii Rchb.f. ex Withner
Laelia mantiqueirae Pabst ex Zappi
Laelia milleri Blumensch. ex Pabst
Laelia mixta Hoehne
Laelia perrinii (Lindl.) Bateman
Laelia pfisteri Pabst & Sènghas
Laelia pumila (Hook.) Rchb.f.
Laelia purpurata Lindl.
Laelia reginae Pabst
Laelia sanguiloba Withner
Laelia sincorana Schltr.
Laelia spectabilis (Paxton) Withner
Laelia tenebrosa (Gower) Rolfe
Laelia tereticaulis Hoehne ex Ruschi
Laelia virens Lindl.
Laelia xanthina Lindl. ex Hook.
Phragmipedium klotzschianum (Rchb.f.) Rolfe
Phragmipedium lindleyanum (Lindl.) Rolfe
Phragmipedium sargentianum (Rolfe) Rolfe
Phragmipedium vittatum (Vell.) Rolfe
Sophronitella violacea (Lindl.) Schltr.
Sophronitis acuensis Fowlie
Sophronitis bicolor F.Miranda
Sophronitis brevipedunculata (Cogn.) Fowlie
Sophronitis cernua (Lindl.) Lindl.
Sophronitis coccinea (Lindl.) Rchb.f.
Sophronitis mantiqueirae (Fowlie) Fowlie
Sophronitis pterocarpa Lindl. & Paxt.
Sophronitis pygmaea (Pabst) Withner
Sophronitis wittigiana Barb. Rodr.

BRUNEI DARUSSALAM

Phalaenopsis amabilis (L.) Blume
Phalaenopsis corningiana Rchb.f.
Phalaenopsis cornucervi (Breda) Blume & Rchb.f.
Phalaenopsis fuscata Rchb.f.
Phalaenopsis gigantea J.J.Sm.
Phalaenopsis lamelligera Sweet
Phalaenopsis maculata Rchb.f.
Phalaenopsis sumatrana Korth. & Rchb.f.

BULGARIA

Cypripedium calceolus L.

CAMBODIA

Paphiopedilum appletonianum (Gower) Rolfe
Paphiopedilum callosum (Rchb.f.) Stein

CAMBODIA (continued)

Paphiopedilum concolor (Lindl.) Pfitzer

CANADA

Cypripedium acaule Aiton
Cypripedium arietinum R.Br.
Cypripedium calceolus L.
Cypripedium candidum Muhl. ex Willd.
Cypripedium × columbianum Sheviak
Cypripedium fasciculatum Kellogg ex S.Watson
Cypripedium montanum Douglas ex Lindl.
Cypripedium parviflorum E.Salisb.
Cypripedium parviflorum var. **pubescens** (Willd.) Knight
Cypripedium passerinum Richardson
Cypripedium reginae Walter

CHINA

Cypripedium × barbeyi Camus
Cypripedium bardolphianum W.W.Sm. & Farrer
Cypripedium bardolphianum var. **zhongdianense** S.C.Chen
Cypripedium calceolus L.
Cypripedium debile Rchb.f.
Cypripedium elegans Rchb.f.
Cypripedium fargesii Franch.
Cypripedium farreri W.W.Sm.
Cypripedium fasciolatum Franch.
Cypripedium flavum Hunt & Summerh.
Cypripedium forrestii P.J.Cribb
Cypripedium franchetii Rolfe
Cypripedium guttatum Sw.
Cypripedium henryi Rolfe
Cypripedium himalaicum Rolfe ex Hemsl.
Cypripedium japonicum Thunb.
Cypripedium lichiangense P.J.Cribb & S.C.Chen
Cypripedium macranthon Sw.
Cypripedium macranthon var. **rebunense** Miyabe & Kudo
Cypripedium margaritaceum Franch.
Cypripedium micranthum Franch.
Cypripedium palangshanense Tang & Wang
Cypripedium plectrochilum Franch.
Cypripedium shanxiense S.C.Chen
Cypripedium smithii Schltr.
Cypripedium subtropicum S.C.Chen & K.Y.Lang
Cypripedium tibeticum King ex Hemsl.
Cypripedium wardii Rolfe
Cypripedium wumengense S.C.Chen
Cypripedium yunnanense Franch.
Paphiopedilum appletonianum (Gower) Rolfe
Paphiopedilum armeniacum S.C.Chen & F.Y.Liu
Paphiopedilum barbatum (Lindl.) Pfitzer
Paphiopedilum barbigerum T.Tang & F.T.Wang
Paphiopedilum bellatulum (Rchb.f.) Stein
Paphiopedilum callosum (Rchb.f.) Stein

CHINA (continued)

Paphiopedilum callosum var. **sublaeve** (Rchb.f.) P.J.Cribb
Paphiopedilum charlesworthii (Rolfe) Pfitzer
Paphiopedilum concolor (Lindl.) Pfitzer
Pleione × confusa P.J.Cribb & C.Z.Tang
Paphiopedilum delenatii Guillaumin
Paphiopedilum dianthum T.Tang & F.T.Wang
Paphiopedilum emersonii Koop. & P.J.Cribb
Paphiopedilum exul (Ridl.) Rolfe
Paphiopedilum godefroyae (God.-Leb.) Stein
Paphiopedilum godefroyae var. **leucochilum** (Masters) Hallier
Paphiopedilum gratrixianum (Masters) Guillaumin
Paphiopedilum henryanum Braem
Paphiopedilum hirsutissimum (Lindl. ex Hook.) Stein
Paphiopedilum hirsutissimum var. **esquirolei** (Schltr.) P.J.Cribb
Paphiopedilum malipoense Chen & Z.H.Tsi
Paphiopedilum micranthum T.Tang & C.W.Wang
Paphiopedilum niveum (Rchb.f.) Stein
Paphiopedilum parishii (Rchb.f.) Stein
Paphiopedilum purpuratum (Lindl.) Stein
Paphiopedilum spicerianum (Rchb.f. ex Masters & T.Moore) Pfitzer
Paphiopedilum sukhakulii Schoser & Senghas
Paphiopedilum tigrinum Koop. & N.Haseg.
Paphiopedilum villosum (Lindl.) Stein
Paphiopedilum villosum var. **annamense** Rolfe
Paphiopedilum villosum var. **boxallii** (Rchb.f.) Pfitzer
Paphiopedilum wardii Summerh.
Phalaenopsis stobartiana Rchb.f.
Phalaenopsis wilsonii Rolfe
Pleione albiflora P.J.Cribb & C.Z.Tang
Pleione amoena Schltr.
Pleione bulbocodioides (Franch.) Rolfe
Pleione chunii C.L.Tso
Pleione formosana Hayata
Pleione forrestii Schltr.
Pleione grandiflora (Rolfe) Rolfe
Pleione hookeriana (Lindl.) B.S.Williams
Pleione humilis (J.E.Sm.) D.Don
Pleione × kohlsii Braem
Pleione limprichtii Schltr.
Pleione maculata (Lindl.) Lindl.
Pleione praecox (J.E.Sm.) D.Don
Pleione saxicola T.Tang & Wang ex S.C.Chen
Pleione scopulorum W.W.Sm.
Pleione speciosa Ames & Schltr.
Pleione yunnanensis (Rolfe) Rolfe

CHINA (TAIWAN)

Cypripedium debile Rchb.f.
Cypripedium formosanum Hayata
Cypripedium macranthum Sw.
Cypripedium segawai Masam.
Phalaenopsis aphrodite Rchb.f.
Phalaenopsis equestris (Schauer) Rchb.f.

Part III: Country Checklist

CHINA (TIBET)

Cypripedium cordigerum D.Don
Cypripedium flavum Hunt & Summerh.
Cypripedium macranthum var. **rebunense** Miyabe & Kudo

COLOMBIA

Cattleya aurea (T.Moore,R.Warner & B.S.Williams) Rodigas
Cattleya candida (Kunth) Lehm.
Cattleya deckeri Klotzsch
Cattleya gaskelliana (hort. ex N.E.Br.) Withner
Cattleya hardyana Hardy ex B.S.Williams
Cattleya maxima Lindl.
Cattleya mendelii Backh.f. ex B.S.Williams
Cattleya schroderae (Rchb.f.) Sander
Cattleya trianaei Linden & Rchb.f.
Cattleya violacea (Kunth) Rolfe
Cattleya warscewiczii Rchb.f.
Phragmipedium hirtzii Dodson
Phragmipedium lindenii (Lindl.) Dressler & N.H.Williams
Phragmipedium longifolium (Rchb.f. & Warsz.) Rolfe
Phragmipedium schlimii (Linden & Rchb.f.) Rolfe
Phragmipedium wallisii (Rchb.f.) Garay

COSTA RICA

Cattleya deckeri Klotzsch
Cattleya dowiana Bateman
Cattleya skinneri Bateman
Laelia rubescens Lindl.
Phragmipedium caudatum (Lindl.) Rolfe
Phragmipedium longifolium (Rchb.f. & Warsz.) Rolfe

CZECH REPUBLIC

Cypripedium calceolus L.

DENMARK

Cypripedium calceolus L.

ECUADOR

Cattleya iricolor Rchb.f.
Cattleya luteola Lindl.
Cattleya maxima Lindl.
Cattleya violacea (Kunth) Rolfe
Phragmipedium besseae Dodson & E.Kuhn
Phragmipedium boissierianum (Rchb.f.) Rolfe
Phragmipedium caudatum (Lindl.) Rolfe
Phragmipedium hirtzii Dodson
Phragmipedium lindenii (Lindl.) Dressler & N.H.Williams
Phragmipedium longifolium (Rchb.f. & Warsz.) Rolfe
Phragmipedium pearcei (Rchb.f.) Rauh & Senghas
Phragmipedium portillae Gruss & Roeth

ECUADOR (continued)

Phragmipedium wallisii (Rchb.f.) Garay

EL SALVADOR

Cattleya aurantiaca (Bateman ex Lindl.) P.N.Don
Cattleya deckeri Klotzsch
Laelia rubescens Lindl.

FINLAND

Cypripedium calceolus L.

FRANCE

Cypripedium calceolus L.

FRENCH GUIANA

Phragmipedium lindleyanum (Lindl.) Rolfe

GERMANY

Cypripedium calceolus L.

GREECE

Cypripedium calceolus L.

GUATEMALA

Cattleya aurantiaca (Bateman ex Lindl.) P.N.Don
Cattleya bowringiana O'Brien
Cattleya deckeri Klotzsch
Cattleya guatemalensis T.Moore
Cattleya skinneri Bateman
Cypripedium irapeanum La Llave & Lex.
Laelia rubescens Lindl.
Phragmipedium caudatum (Lindl.) Rolfe

GUYANA

Cattleya jenmanii Rolfe
Cattleya lawrenceana Rchb.f.
Cattleya violacea (Kunth) Rolfe
Phragmipedium klotzschianum (Rchb.f.) Rolfe
Phragmipedium lindleyanum (Lindl.) Rolfe

HONDURAS

Cattleya aurantiaca (Bateman ex Lindl.) P.N.Don
Cattleya deckeri Klotzsch
Cattleya skinneri Bateman
Cypripedium irapeanum La Llave & Lex.

HONDURAS (continued)

Laelia rubescens Lindl.

HONG KONG

Paphiopedilum purpuratum (Lindl.) Stein

HUNGARY

Cypripedium calceolus L.

INDIA

Cypripedium cordigerum D.Don
Cypripedium elegans Rchb.f.
Cypripedium guttatum Sw.
Cypripedium himalaicum Rolfe ex Hemsl.
Paphiopedilum druryi (Bedd.) Stein
Paphiopedilum fairrieanum (Lindl.) Stein
Paphiopedilum hirsutissimum (Lindl. ex Hook.) Stein
Paphiopedilum insigne (W.Wall ex Lindl.) Pfitzer
Paphiopedilum spicerianum (Rchb.f. ex Masters & T.Moore) Pfitzer
Paphiopedilum venustum (Wall. ex Sims) Pfitzer ex Stein
Paphiopedilum villosum (Lindl.) Stein
Phalaenopsis mysorensis Saldanha
Pleione hookeriana (Lindl.) B.S.Williams
Pleione humilis (J.E.Sm.) D.Don
Pleione × lagenaria Lindl.
Pleione maculata (Lindl.) Lindl.
Pleione praecox (J.E.Sm.) D.Don

INDONESIA

Paphiopedilum bullenianum (Rchb.f.) Pfitzer
Paphiopedilum bullenianum var. **celebesense** (Fowlie & Birk) P.J.Cribb
Paphiopedilum glanduliferum (Blume) Stein
Paphiopedilum glanduliferum var. **wilhelminae** (L.O.Williams) P.J.Cribb
Paphiopedilum glaucophyllum J.J.Sm.
Paphiopedilum glaucophyllum var. **moquetteanum** J.J.Sm.
Paphiopedilum hookerae (Rchb.f.) Stein
Paphiopedilum javanicum (Reinw. ex Lindl.) Pfitzer
Paphiopedilum kolopakingii Fowlie
Paphiopedilum liemianum (Fowlie) Karas. & K.Saito
Paphiopedilum lowii (Lindl.) Stein
Paphiopedilum lowii var. **richardianum** (Asher & Beaman) Gruss
Paphiopedilum mastersianum (Rchb.f.) Stein
Paphiopedilum mohrianum Braem
Paphiopedilum papuanum (Ridl.) Ridl.
Paphiopedilum primulinum M.W.Wood & Taylor
Paphiopedilum primulinum var. **purpurascens** (M.W.Wood) P.J.Cribb
Paphiopedilum sangii Braem
Paphiopedilum schoseri Braem & H.Mohr
Paphiopedilum supardii Braem & Loeb
Paphiopedilum superbiens (Rchb.f.) Stein
Paphiopedilum superbiens var. **curtisii** (Rchb.f.) G.J.Braem

INDONESIA (continued)

Paphiopedilum tonsum (Rchb.f.) Stein
Paphiopedilum tonsum var. **braemii** (Mohr) Gruss
Paphiopedilum victoria-mariae (Sander ex Mast.) Rolfe
Paphiopedilum victoria-regina (Sander) M.W.Wood
Paphiopedilum violascens Schltr.
Paraphalaenopsis denevei (J.J.Sm.) A.D.Hawkes
Paraphalaenopsis laycockii (M.R.Hend.) A.D.Hawkes
Paraphalaenopsis serpentilingua (J.J.Sm.) A.D.Hawkes
Paraphalaenopsis × thorntonii (Holttum) A.D.Hawkes
Phalaenopsis amabilis (L.) Blume
Phalaenopsis amabilis var. **moluccana** Schltr.
Phalaenopsis amboinensis J.J.Sm.
Phalaenopsis celebensis Sweet
Phalaenopsis corningiana Rchb.f.
Phalaenopsis cornucervi (Breda) Blume & Rchb.f.
Phalaenopsis cornucervi var. **picta** (Hassk.) Sweet
Phalaenopsis fimbriata J.J.Sm.
Phalaenopsis fimbriata var. **sumatrana** J.J.Sm.
Phalaenopsis floresensis Fowlie
Phalaenopsis fuscata Rchb.f.
Phalaenopsis × gersenii (Teijsm. & Binn.) Rolfe
Phalaenopsis gigantea J.J.Sm.
Phalaenopsis inscriptiosinensis Fowlie
Phalaenopsis javanica J.J.Sm.
Phalaenopsis lamelligera Sweet
Phalaenopsis maculata Rchb.f.
Phalaenopsis mariae Burb. ex R.Warner & B.S.Williams
Phalaenopsis modesta J.J.Sm.
Phalaenopsis modesta var. **bella** Gruss & Roellke
Phalaenopsis pantherina Rchb.f.
Phalaenopsis robinsonii J.J.Sm.
Phalaenopsis rosenstromii Bailey
Phalaenopsis speciosa Rchb.f.
Phalaenopsis speciosa var. **christiana** Rchb.f.
Phalaenopsis speciosa var. **imperatrix** Rchb.f.
Phalaenopsis sumatrana Korth. & Rchb.f.
Phalaenopsis sumatrana var. **alba** G.Wilson
Phalaenopsis sumatrana var. **paucivittata** Rchb.f.
Phalaenopsis tetraspis Rchb.f.
Phalaenopsis venosa Shim & Fowlie
Phalaenopsis violacea Witte
Phalaenopsis violacea var. **alba** Teijsm. & Binn.
Phalaenopsis violacea var. **bowringiana** Rchb.f.
Phalaenopsis viridis J.J.Sm.

ITALY

Cypripedium calceolus L.

JAPAN

Cypripedium calceolus L.
Cypripedium debile Rchb.f.
Cypripedium guttatum Sw.

JAPAN (continued)

Cypripedium japonicum Thunb.
Cypripedium macranthon Sw.
Cypripedium macranthon var. **rebunense** Miyabe & Kudo
Cypripedium shanxiense S.C.Chen
Cypripedium yatabeanum Makino

KOREA

Cypripedium × barbeyi Camus
Cypripedium calceolus L.
Cypripedium guttatum Sw.
Cypripedium japonicum Thunb.
Cypripedium macranthon Sw.

LAO PEOPLE'S DEMOCRATIC REPUBLIC

Paphiopedilum appletonianum (Gower) Rolfe
Paphiopedilum bellatulum (Rchb.f.) Stein
Paphiopedilum callosum (Rchb.f.) Stein
Paphiopedilum concolor (Lindl.) Pfitzer
Paphiopedilum gratrixianum (Masters) Guillaumin
Paphiopedilum villosum var. **annamense** Rolfe
Phalaenopsis gibbosa Sweet
Pleione bulbocodioides (Franch.) Rolfe
Pleione hookeriana (Lindl.) B.S.Williams
Pleione praecox (J.E.Sm.) D.Don

LITHUANIA

Cypripedium calceolus L.

LUXEMBOURG

Cypripedium calceolus L.

MALAYSIA

Paphiopedilum barbatum (Lindl.) Pfitzer
Paphiopedilum bullenianum (Rchb.f.) Pfitzer
Paphiopedilum callosum var. **sublaeve** (Rchb.f.) P.J.Cribb
Paphiopedilum dayanum (Lindl.) Stein
Paphiopedilum hookerae (Rchb.f.) Stein
Paphiopedilum hookerae var. **volonteanum** (Sander ex Rolfe) Kerch.
Paphiopedilum javanicum (Reinw. ex Lindl.) Pfitzer
Paphiopedilum javanicum var. **virens** (Rchb.f.) Stein
Paphiopedilum lawrenceanum (Rchb.f.) Pfitzer
Paphiopedilum lowii (Lindl.) Stein
Paphiopedilum niveum (Rchb.f.) Stein
Paphiopedilum philippinense (Rchb.f.) Stein
Paphiopedilum rothschildianum (Rchb.f.) Stein
Paphiopedilum sanderianum (Rchb.f.) Stein
Paphiopedilum schoseri Braem & H.Mohr
Paphiopedilum stonei (Hook.) Stein
Paraphalaenopsis labukensis Shim, Lamb & Chan

MALAYSIA (continued)

Paraphalaenopsis serpentilingua (J.J.Sm.) A.D.Hawkes
Paraphalaenopsis × thorntonii (Holttum) A.D.Hawkes
Phalaenopsis amabilis (L.) Blume
Phalaenopsis appendiculata C.E.Carr
Phalaenopsis cochlearis Holttum
Phalaenopsis corningiana Rchb.f.
Phalaenopsis cornucervi (Breda) Blume & Rchb.f.
Phalaenopsis fimbriata J.J.Sm.
Phalaenopsis fuscata Rchb.f.
Phalaenopsis × gersenii (Teijsm. & Binn.) Rolfe
Phalaenopsis gigantea J.J.Sm.
Phalaenopsis kunstleri Hook.f.
Phalaenopsis lamelligera Sweet
Phalaenopsis maculata Rchb.f.
Phalaenopsis mariae Burb. ex R.Warner & B.S.Williams
Phalaenopsis modesta J.J.Sm.
Phalaenopsis modesta var. **bella** Gruss & Roellke
Phalaenopsis pantherina Rchb.f.
Phalaenopsis sumatrana Korth. & Rchb.f.
Phalaenopsis × valentinii Rchb.f.
Phalaenopsis violacea Witte
Phalaenopsis violacea var. **alba** Teijsm. & Binn.
Phalaenopsis violacea var. **murtoniana** Rchb.f.

MEXICO

Cattleya aurantiaca (Bateman ex Lindl.) P.N.Don
Cypripedium dickinsonianum Hágsater
Cypripedium irapeanum La Llave & Lex.
Cypripedium molle Lindl.
Laelia albida Bateman ex Lindl.
Laelia anceps Lindl.
Laelia autumnalis (La Llave & Lex.) Lindl.
Laelia bancalarii R.Gonzalez & Hágsater
Laelia furfuracea Lindl.
Laelia gouldiana Rchb.f.
Laelia rubescens Lindl.
Laelia speciosa (Kunth) Schltr.
Phragmipedium exstaminodium Castano, Hágsater & E.Aguirre
Phragmipedium xerophyticum J.C.Soto, Salazar & Hágsater

MONGOLIA

Cypripedium calceolus L.

MYANMAR

Cypripedium lichiangense P.J.Cribb & S.C.Chen
Paphiopedilum bellatulum (Rchb.f.) Stein
Paphiopedilum charlesworthii (Rolfe) Pfitzer
Paphiopedilum concolor (Lindl.) Pfitzer
Paphiopedilum hirsutissimum (Lindl. ex Hook.) Stein
Paphiopedilum parishii (Rchb.f.) Stein
Paphiopedilum spicerianum (Rchb.f. ex Masters & T.Moore) Pfitzer

MYANMAR (continued)

Paphiopedilum villosum (Lindl.) Stein
Paphiopedilum villosum var. **boxallii** (Rchb.f.) Pfitzer
Paphiopedilum wardii Summerh.
Phalaenopsis cornucervi (Breda) Blume & Rchb.f.
Phalaenopsis kunstleri Hook.f.
Phalaenopsis lobbii (Rchb.f.) Sweet
Phalaenopsis lowii Rchb.f.
Phalaenopsis parishii Rchb.f.
Phalaenopsis sumatrana Korth. & Rchb.f.
Phalaenopsis tetraspis Rchb.f.
Pleione albiflora P.J.Cribb & C.Z.Tang
Pleione forrestii Schltr.
Pleione hookeriana (Lindl.) B.S.Williams
Pleione humilis (J.E.Sm.) D.Don
Pleione maculata (Lindl.) Lindl.
Pleione praecox (J.E.Sm.) D.Don
Pleione scopulorum W.W.Sm.
Pleione yunnanensis (Rolfe) Rolfe

NEPAL

Cypripedium cordigerum D.Don
Cypripedium elegans Rchb.f.
Cypripedium guttatum Sw.
Cypripedium himalaicum Rolfe ex Hemsl.
Paphiopedilum insigne (W.Wall ex Lindl.) Pfitzer
Paphiopedilum venustum (Wall. ex Sims) Pfitzer ex Stein
Pleione coronaria P.J.Cribb & C.Z.Tang
Pleione hookeriana (Lindl.) B.S.Williams
Pleione humilis (J.E.Sm.) D.Don
Pleione praecox (J.E.Sm.) D.Don

NICARAGUA

Cattleya deckeri Klotzsch
Cattleya skinneri Bateman
Laelia rubescens Lindl.

NORWAY

Cypripedium calceolus L.

PAKISTAN

Cypripedium cordigerum D.Don

PANAMA

Cattleya skinneri Bateman
Laelia rubescens Lindl.
Phragmipedium caudatum (Lindl.) Rolfe
Phragmipedium longifolium (Rchb.f. & Warsz.) Rolfe

PAPUA NEW GUINEA

Paphiopedilum glanduliferum (Blume) Stein
Paphiopedilum glanduliferum var. **wilhelminae** (L.O.Williams) P.J.Cribb
Paphiopedilum papuanum (Ridl.) Ridl.
Paphiopedilum violascens Schltr.
Phalaenopsis rosenstromii Bailey

PAPUA NEW GUINEA (BOUGAINVILLE)

Paphiopedilum bougainvilleanum Fowlie
Paphiopedilum wentworthianum Schoser & Fowlie

PARAGUAY

Sophronitis cernua (Lindl.) Lindl.
Sophronitis pterocarpa Lindl. & Paxt.

PERU

Cattleya luteola Lindl.
Cattleya maxima Lindl.
Cattleya mooreana Withner, Allison & Guenard
Cattleya rex O'Brien
Cattleya violacea (Kunth) Rolfe
Phragmipedium besseae Dodson & E.Kuhn
Phragmipedium boissierianum (Rchb.f.) Rolfe
Phragmipedium caudatum (Lindl.) Rolfe
Phragmipedium pearcei (Rchb.f.) Rauh & Senghas
Phragmipedium richteri Roeth & Gruss

PHILIPPINES

Paphiopedilum acmodontum Schoser ex M.W.Wood
Paphiopedilum adductum Asher
Paphiopedilum argus (Rchb.f.) Stein
Paphiopedilum ciliolare (Rchb.f.) Stein
Paphiopedilum fowliei Birk
Paphiopedilum haynaldianum (Rchb.f.) Stein
Paphiopedilum hennisianum (M.W.Wood) Fowlie
Paphiopedilum philippinense (Rchb.f.) Stein
Paphiopedilum philippinense var. **roebelenii** (Veitch) P.J.Cribb
Paphiopedilum randsii Fowlie
Paphiopedilum urbanianum Fowlie
Phalaenopsis amabilis (L.) Blume
Phalaenopsis aphrodite Rchb.f.
Phalaenopsis bastianii Gruss & Roellke
Phalaenopsis cornucervi (Breda) Blume & Rchb.f.
Phalaenopsis equestris (Schauer) Rchb.f.
Phalaenopsis equestris var. **alba** hort.
Phalaenopsis equestris var. **leucaspis** Rchb.f.
Phalaenopsis equestris var. **leucotanthe** Rchb.f. ex God.-Leb.
Phalaenopsis equestris var. **rosea** Valmayor & Tiu
Phalaenopsis fasciata Rchb.f.
Phalaenopsis fuscata Rchb.f.
Phalaenopsis hieroglyphica (Rchb.f.) Sweet

PHILIPPINES (continued)

Phalaenopsis × intermedia Lindl.
Phalaenopsis × leucorrhoda Rchb.f.
Phalaenopsis lindenii Loher
Phalaenopsis lueddemanniana Rchb.f.
Phalaenopsis lueddemanniana var. **delicata** Rchb.f.
Phalaenopsis lueddemanniana var. **ochracea** Rchb.f.
Phalaenopsis mariae Burb. ex R.Warner & B.S.Williams
Phalaenopsis micholitzii Rolfe
Phalaenopsis pallens (Lindl.) Rchb.f.
Phalaenopsis pallens var. **alba** (Ames & Quisumb.) Sweet
Phalaenopsis pallens var. **denticulata** (Rchb.f.) Sweet
Phalaenopsis philippinensis Golamco ex Fowlie & Tang
Phalaenopsis pulchra (Rchb.f.) Sweet
Phalaenopsis reichenbachiana Rchb.f. & Sander
Phalaenopsis sanderiana Rchb.f.
Phalaenopsis sanderiana var. **alba** (J.Veitch) Stein
Phalaenopsis sanderiana var. **marmorata** Rchb.f.
Phalaenopsis schilleriana Rchb.f.
Phalaenopsis schilleriana var. **immaculata** Rchb.f.
Phalaenopsis schilleriana var. **purpurea** O'Brien
Phalaenopsis schilleriana var. **splendens** R.Warner
Phalaenopsis stuartiana Rchb.f.
Phalaenopsis × veitchiana Rchb.f.
Phalaenopsis venosa Shim & Fowlie

POLAND

Cypripedium calceolus L.

ROMANIA

Cypripedium calceolus L.

RUSSIAN FEDERATION

Cypripedium × barbeyi Camus
Cypripedium calceolus L.
Cypripedium guttatum Sw.
Cypripedium macranthon Sw.
Cypripedium macranthon var. **rebunense** Miyabe & Kudo
Cypripedium shanxiense S.C.Chen
Cypripedium yatabeanum Makino

SOLOMON ISLANDS

Paphiopedilum wentworthianum Schoser & Fowlie

SLOVAK REPUBLIC

Cypripedium calceolus L.

SPAIN

Cypripedium calceolus L.

SURINAME

Phragmipedium lindleyanum (Lindl.) Rolfe

SWEDEN

Cypripedium calceolus L.

SWITZERLAND

Cypripedium calceolus L.

THAILAND

Paphiopedilum appletonianum (Gower) Rolfe
Paphiopedilum barbatum (Lindl.) Pfitzer
Paphiopedilum bellatulum (Rchb.f.) Stein
Paphiopedilum callosum (Rchb.f.) Stein
Paphiopedilum callosum var. **sublaeve** (Rchb.f.) P.J.Cribb
Paphiopedilum charlesworthii (Rolfe) Pfitzer
Paphiopedilum concolor (Lindl.) Pfitzer
Paphiopedilum exul (Ridl.) Rolfe
Paphiopedilum godefroyae (God.-Leb.) Stein
Paphiopedilum godefroyae var. **leucochilum** (Masters) Hallier
Paphiopedilum hirsutissimum var. **esquirolei** (Schltr.) P.J.Cribb
Paphiopedilum niveum (Rchb.f.) Stein
Paphiopedilum parishii (Rchb.f.) Stein
Paphiopedilum sukhakulii Schoser & Senghas
Paphiopedilum villosum (Lindl.) Stein
Phalaenopsis cornucervi (Breda) Blume & Rchb.f.
Phalaenopsis parishii Rchb.f.
Phalaenopsis sumatrana Korth. & Rchb.f.
Phalaenopsis thalebanii Seidenfaden
Pleione hookeriana (Lindl.) B.S.Williams
Pleione maculata (Lindl.) Lindl.
Pleione praecox (J.E.Sm.) D.Don

TRINIDAD & TOBAGO

Cattleya deckeri Klotzsch

UNITED KINGDOM

Cypripedium calceolus L.

USA

Cypripedium acaule Aiton
Cypripedium × andrewsii Fuller
Cypripedium arietinum R.Br.
Cypripedium calceolus L.
Cypripedium californicum A.Gray
Cypripedium candidum Muhl. ex Willd.
Cypripedium × columbianum Sheviak
Cypripedium fasciculatum Kellogg ex S.Watson
Cypripedium guttatum Sw.

USA (continued)

Cypripedium kentuckiense C.F.Reed
Cypripedium montanum Douglas ex Lindl.
Cypripedium parviflorum E.Salisb.
Cypripedium parviflorum var. **pubescens** (Willd.) Knight
Cypripedium passerinum Richardson
Cypripedium reginae Walter
Cypripedium yatabeanum Makino

VENEZUELA

Cattleya deckeri Klotzsch
Cattleya gaskelliana (hort. ex N.E.Br.) Withner
Cattleya jenmanii Rolfe
Cattleya lawrenceana Rchb.f.
Cattleya lueddemanniana Rchb.f.
Cattleya maxima Lindl.
Cattleya mossiae Hook.
Cattleya percivaliana (Rchb.f.) O'Brien
Cattleya violacea (Kunth) Rolfe
Phragmipedium klotzschianum (Rchb.f.) Rolfe
Phragmipedium lindleyanum (Lindl.) Rolfe
Phragmipedium lindleyanum var. **kaieteurum** (N.E.Br.) Rchb.f. ex Pfitzer

VIET NAM

Paphiopedilum appletonianum (Gower) Rolfe
Paphiopedilum callosum (Rchb.f.) Stein
Paphiopedilum concolor (Lindl.) Pfitzer
Paphiopedilum delenatii Guillaumin
Paphiopedilum gratrixianum (Masters) Guillaumin
Paphiopedilum hirsutissimum var. **esquirolei** (Schltr.) P.J.Cribb
Paphiopedilum henryanum Braem
Paphiopedilum malipoense Chen & Z.H.Tsi
Paphiopedilum micranthum T.Tang & C.W.Wang
Paphiopedilum purpuratum (Lindl.) Stein
Paphiopedilum villosum var. **annamense** Rolfe
Phalaenopsis gibbosa Sweet
Phalaenopsis lobbii (Rchb.f.) Sweet
Phalaenopsis mannii Rchb.f.
Phalaenopsis sumatrana Korth. & Rchb.f.

YUGOSLAVIA (FORMER)

Cypripedium calceolus L.

IUCN RED LIST CATEGORIES

Prepared by the

IUCN Species Survival Commission

As approved by the
40th Meeting of the IUCN Council
Gland, Switzerland

30 November 1994

IUCN RED LIST CATEGORIES

I) Introduction

1. The threatened species categories now used in Red Data Books and Red Lists have been in place, with some modification, for almost 30 years. Since their introduction these categories have become widely recognised internationally, and they are now used in a whole range of publications and listings, produced by IUCN as well as by numerous governmental and non-governmental organisations. The Red Data Book categories provide an easily and widely understood method for highlighting those species under higher extinction risk, so as to focus attention on conservation measures designed to protect them.

2. The need to revise the categories has been recognised for some time. In 1984, the SSC held a symposium, 'The Road to Extinction' (Fitter & Fitter 1987), which examined the issues in some detail, and at which a number of options were considered for the revised system. However, no single proposal resulted. The current phase of development began in 1989 with a request from the SSC Steering Committee to develop a new approach that would provide the conservation community with useful information for action planning.

In this document, proposals for new definitions for Red List categories are presented. The general aim of the new system is to provide an explicit, objective framework for the classification of species according to their extinction risk.

The revision has several specific aims:

- to provide a system that can be applied consistently by different people;

- to improve the objectivity by providing those using the criteria with clear guidance on how to evaluate different factors which affect risk of extinction;

- to provide a system which will facilitate comparisons across widely different taxa;

- to give people using threatened species lists a better understanding of how individual species were classified.

3. The proposals presented in this document result from a continuing process of drafting, consultation and validation. It was clear that the production of a large number of draft proposals led to some confusion, especially as each draft has been used for classifying some set of species for conservation purposes. To clarify matters, and to open the way for modifications as and when they became necessary, a system for version numbering was applied as follows:

Version 1.0: Mace & Lande (1991)
The first paper discussing a new basis for the categories, and presenting numerical criteria especially relevant for large vertebrates.

IUCN Red List Categories

Version 2.0: Mace *et al.* (1992)
A major revision of Version 1.0, including numerical criteria appropriate to all organisms and introducing the non-threatened categories.

Version 2.1: IUCN (1993)
Following an extensive consultation process within SSC, a number of changes were made to the details of the criteria, and fuller explanation of basic principles was included. A more explicit structure clarified the significance of the non-threatened categories.

Version 2.2: Mace & Stuart (1994)
Following further comments received and additional validation exercises, some minor changes to the criteria were made. In addition, the Susceptible category present in Versions 2.0 and 2.1 was subsumed into the Vulnerable category. A precautionary application of the system was emphasised.

Final Version
This final document, which incorporates changes as a result of comments from IUCN members, was adopted by the IUCN Council in December 1994.

All future taxon lists including categorisations should be based on this version, and not the previous ones.

4. In the rest of this document the proposed system is outlined in several sections. The Preamble presents some basic information about the context and structure of the proposal, and the procedures that are to be followed in applying the definitions to species. This is followed by a section giving definitions of terms used. Finally the definitions are presented, followed by the quantitative criteria used for classification within the threatened categories. It is important for the effective functioning of the new system that all sections are read and understood, and the guidelines followed.

References:

Fitter, R., and M. Fitter, ed. (1987) The Road to Extinction. Gland, Switzerland: IUCN.

IUCN. (1993) Draft IUCN Red List Categories. Gland, Switzerland: IUCN.

Mace, G. M. *et al.* (1992) "The development of new criteria for listing species on the IUCN Red List." Species 19: 16-22.

Mace, G. M., and R. Lande. (1991) "Assessing extinction threats: toward a reevaluation of IUCN threatened species categories." Conserv. Biol. 5.2: 148-157.

Mace, G. M. & S. N. Stuart. (1994) "Draft IUCN Red List Categories, Version 2.2". Species 21-22: 13-24.

II) Preamble

The following points present important information on the use and interpretation of the categories (= Critically Endangered, Endangered, etc.), criteria (= A to E), and sub-criteria (= a,b etc., i,ii etc.):

1. **Taxonomic level and scope of the categorisation process**

The criteria can be applied to any taxonomic unit at or below the species level. The term 'taxon' in the following notes, definitions and criteria is used for convenience, and may represent species or lower taxonomic levels, including forms that are not yet formally described. There is a sufficient range among the different criteria to enable the appropriate listing of taxa from the complete taxonomic spectrum, with the exception of micro-organisms. The criteria may also be applied within any specified geographical or political area although in such cases special notice should be taken of point 11 below. In presenting the results of applying the criteria, the taxonomic unit and area under consideration should be made explicit. The categorisation process should only be applied to wild populations inside their natural range, and to populations resulting from benign introductions (defined in the draft IUCN Guidelines for Re-introductions as "..an attempt to establish a species, for the purpose of conservation, outside its recorded distribution, but within an appropriate habitat and eco-geographical area").

2. **Nature of the categories**

All taxa listed as Critically Endangered qualify for Vulnerable and Endangered, and all listed as Endangered qualify for Vulnerable. Together these categories are described as 'threatened'. The threatened species categories form a part of the overall scheme. It will be possible to place all taxa into one of the categories (see Figure 1).

Figure 1: Structure of the Categories

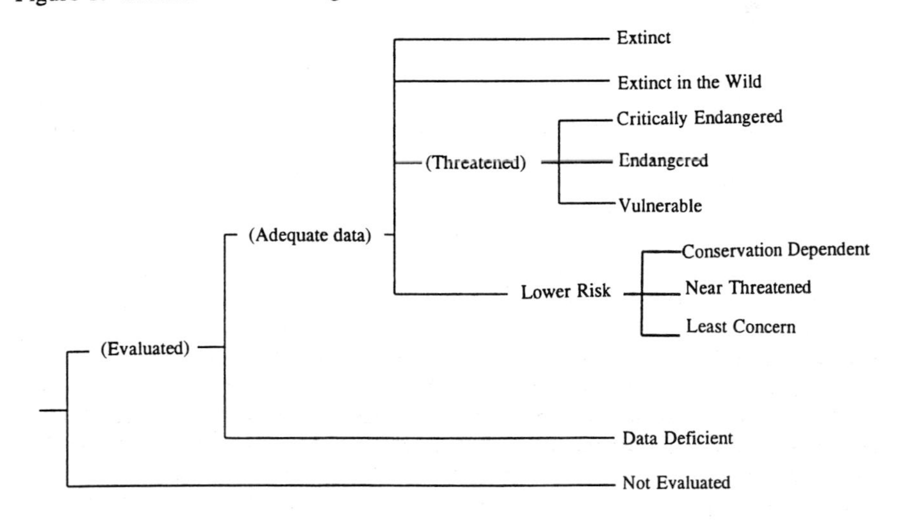

3. **Role of the different criteria**
For listing as Critically Endangered, Endangered or Vulnerable there is a range of quantitative criteria; meeting any one of these criteria qualifies a taxon for listing at that level of threat. Each species should be evaluated against all the criteria. The different criteria (A-E) are derived from a wide review aimed at detecting risk factors across the broad range of organisms and the diverse life histories they exhibit. Even though some criteria will be inappropriate for certain taxa (some taxa will never qualify under these however close to extinction they come), there should be criteria appropriate for assessing threat levels for any taxon (other than micro-organisms). The relevant factor is whether any one criterion is met, not whether all are appropriate or all are met. Because it will never be clear which criteria are appropriate for a particular species in advance, each species should be evaluated against all the criteria, and any criterion met should be listed.

4. **Derivation of quantitative criteria**
The quantitative values presented in the various criteria associated with threatened categories were developed through wide consultation and they are set at what are generally judged to be appropriate levels, even if no formal justification for these values exists. The levels for different criteria within categories were set independently but against a common standard. Some broad consistency between them was sought. However, a given taxon should not be expected to meet all criteria (A-E) in a category; meeting any one criterion is sufficient for listing.

5. **Implications of listing**
Listing in the categories of Not Evaluated and Data Deficient indicates that no assessment of extinction risk has been made, though for different reasons. Until such time as an assessment is made, species listed in these categories should not be treated as if they were non-threatened, and it may be appropriate (especially for Data Deficient forms) to give them the same degree of protection as threatened taxa, at least until their status can be evaluated.

Extinction is assumed here to be a chance process. Thus, a listing in a higher extinction risk category implies a higher expectation of extinction, and over the time-frames specified more taxa listed in a higher category are expected to go extinct than in a lower one (without effective conservation action). However, the persistence of some taxa in high risk categories does not necessarily mean their initial assessments are inaccurate.

6. **Data quality and the importance of inference and projection**
The criteria are clearly quantitative in nature. However, the absence of high quality data should not deter attempts at applying the criteria, as methods involving estimation, inference and projection are emphasised to be acceptable throughout. Inference and projection may be based on extrapolation of current or potential threats into the future (including their rate of change), or of factors related to population abundance or distribution (including dependence on other taxa), so long as these can reasonably be supported. Suspected or inferred patterns in either the recent past, present or near future can be based on any of a series of related factors, and these factors should be specified.

Taxa at risk from threats posed by future events of low probability but with severe consequences (catastrophes) should be identified by the criteria (e.g. small distributions, few locations). Some threats need to be identified particularly early, and appropriate actions taken, because their effects are irreversible, or nearly so (pathogens, invasive organisms, hybridization).

7. **Uncertainty**

The criteria should be applied on the basis of the available evidence on taxon numbers, trend and distribution, making due allowance for statistical and other uncertainties. Given that data are rarely available for the whole range or population of a taxon, it may often be appropriate to use the information that is available to make intelligent inferences about the overall status of the taxon in question. In cases where a wide variation in estimates is found, it is legitimate to apply the precautionary principle and use the estimate (providing it is credible) that leads to listing in the category of highest risk.

Where data are insufficient to assign a category (including Lower Risk), the category of 'Data Deficient' may be assigned. However, it is important to recognise that this category indicates that data are inadequate to determine the degree of threat faced by a taxon, not necessarily that the taxon is poorly known. In cases where there are evident threats to a taxon through, for example, deterioration of its only known habitat, it is important to attempt threatened listing, even though there may be little direct information on the biological status of the taxon itself. The category 'Data Deficient' is not a threatened category, although it indicates a need to obtain more information on a taxon to determine the appropriate listing.

8. **Conservation actions in the listing process**

The criteria for the threatened categories are to be applied to a taxon whatever the level of conservation action affecting it. In cases where it is only conservation action that prevents the taxon from meeting the threatened criteria, the designation of 'Conservation Dependent' is appropriate. It is important to emphasise here that a taxon require conservation action even if it is not listed as threatened.

9. **Documentation**

All taxon lists including categorisation resulting from these criteria should state the criteria and sub-criteria that were met. No listing can be accepted as valid unless at least one criterion is given. If more than one criterion or sub-criterion was met, then each should be listed. However, failure to mention a criterion should not necessarily imply that it was not met. Therefore, if a re-evaluation indicates that the documented criterion is no longer met, this should not result in automatic down-listing. Instead, the taxon should be re-evaluated with respect to all criteria to indicate its status. The factors responsible for triggering the criteria, especially where inference and projection are used, should at least be logged by the evaluator, even if they cannot be included in published lists.

10. **Threats and priorities**

The category of threat is not necessarily sufficient to determine priorities for conservation action. The category of threat simply provides an assessment of the likelihood of extinction under current circumstances, whereas a system for assessing priorities for action will include numerous other factors concerning conservation action such as costs, logistics, chances of success, and even perhaps the taxonomic distinctiveness of the subject.

11. **Use at regional level**

The criteria are most appropriately applied to whole taxa at a global scale, rather than to those units defined by regional or national boundaries. Regionally or nationally based threat categories, which are aimed at including taxa that are threatened at regional or national levels (but not necessarily throughout their global ranges), are best used with two key pieces of information: the global status category for the taxon, and the proportion of the global population or range that occurs within the region or nation. However, if applied at regional or national level it must be recognised that a global category of threat may not be the same as a regional or national category for a particular taxon. For example, taxa classified as Vulnerable on the basis of their global declines in numbers or range might be Lower Risk within a particular region where their populations are stable. Conversely, taxa classified as Lower Risk globally might be Critically Endangered within a particular region where numbers are very small or declining, perhaps only because they are at the margins of their global range. IUCN is still in the process of developing guidelines for the use of national red list categories.

12. **Re-evaluation**

Evaluation of taxa against the criteria should be carried out at appropriate intervals. This is especially important for taxa listed under Near Threatened, or Conservation Dependent, and for threatened species whose status is known or suspected to be deteriorating.

13. **Transfer between categories**

There are rules to govern the movement of taxa between categories. These are as follows: (A) A taxon may be moved from a category of higher threat to a category of lower threat if none of the criteria of the higher category has been met for 5 years or more. (B) If the original classification is found to have been erroneous, the taxon may be transferred to the appropriate category or removed from the threatened categories altogether, without delay (but see Section 9). (C) Transfer from categories of lower to higher risk should be made without delay.

14. **Problems of scale**

Classification based on the sizes of geographic ranges or the patterns of habitat occupancy is complicated by problems of spatial scale. The finer the scale at which the distributions or habitats of taxa are mapped, the smaller will be the area that they are found to occupy. Mapping at finer scales reveals more areas in which the taxon is unrecorded. It is impossible to provide any strict but general rules for mapping taxa or habitats; the most appropriate scale will depend on the taxa in question, and the origin and comprehensiveness of the distributional data. However, the thresholds for some criteria (e.g. Critically Endangered) necessitate mapping at a fine scale.

III) Definitions

1. **Population**

Population is defined as the total number of individuals of the taxon. For functional reasons, primarily owing to differences between life-forms, population numbers are expressed as numbers of mature individuals only. In the case of taxa obligately dependent on other taxa for all or part of their life cycles, biologically appropriate values for the host taxon should be used.

2. Subpopulations

Subpopulations are defined as geographically or otherwise distinct groups in the population between which there is little exchange (typically one successful migrant individual or gamete per year or less).

3. Mature individuals

The number of mature individuals is defined as the number of individuals known, estimated or inferred to be capable of reproduction. When estimating this quantity the following points should be borne in mind:

- Where the population is characterised by natural fluctuations the minimum number should be used.

- This measure is intended to count individuals capable of reproduction and should therefore exclude individuals that are environmentally, behaviourally or otherwise reproductively suppressed in the wild.

- In the case of populations with biased adult or breeding sex ratios it is appropriate to use lower estimates for the number of mature individuals which take this into account (e.g. the estimated effective population size).

- Reproducing units within a clone should be counted as individuals, except where such units are unable to survive alone (e.g. corals).

- In the case of taxa that naturally lose all or a subset of mature individuals at some point in their life cycle, the estimate should be made at the appropriate time, when mature individuals are available for breeding.

4. Generation

Generation may be measured as the average age of parents in the population. This is greater than the age at first breeding, except in taxa where individuals breed only once.

5. Continuing decline

A continuing decline is a recent, current or projected future decline whose causes are not known or not adequately controlled and so is liable to continue unless remedial measures are taken. Natural fluctuations will not normally count as a continuing decline, but an observed decline should not be considered to be part of a natural fluctuation unless there is evidence for this.

6. Reduction

A reduction (criterion A) is a decline in the number of mature individuals of at least the amount (%) stated over the time period (years) specified, although the decline need not still be continuing. A reduction should not be interpreted as part of a natural fluctuation unless there is good evidence for this. Downward trends that are part of natural fluctuations will not normally count as a reduction.

7. **Extreme fluctuations**
Extreme fluctuations occur in a number of taxa where population size or distribution area varies widely, rapidly and frequently, typically with a variation greater than one order of magnitude (i.e., a tenfold increase or decrease).

8. **Severely fragmented**
Severely fragmented refers to the situation where increased extinction risks to the taxon result from the fact that most individuals within a taxon are found in small and relatively isolated subpopulations. These small subpopulations may go extinct, with a reduced probability of recolonisation.

9. **Extent of occurrence**
Extent of occurrence is defined as the area contained within the shortest continuous imaginary boundary which can be drawn to encompass all the known, inferred or projected sites of present occurrence of a taxon, excluding cases of vagrancy. This measure may exclude discontinuities or disjunctions within the overall distributions of taxa (e.g., large areas of obviously unsuitable habitat) (but see 'area of occupancy'). Extent of occurrence can often be measured by a minimum convex polygon (the smallest polygon in which no internal angle exceeds 180 degrees and which contains all the sites of occurrence).

10. **Area of occupancy**
Area of occupancy is defined as the area within its 'extent of occurrence' (see definition) which is occupied by a taxon, excluding cases of vagrancy. The measure reflects the fact that a taxon will not usually occur throughout the area of its extent of occurrence, which may, for example, contain unsuitable habitats. The area of occupancy is the smallest area essential at any stage to the survival of existing populations of a taxon (e.g. colonial nesting sites, feeding sites for migratory taxa). The size of the area of occupancy will be a function of the scale at which it is measured, and should be at a scale appropriate to relevant biological aspects of the taxon. The criteria include values in km^2, and thus to avoid errors in classification, the area of occupancy should be measured on grid squares (or equivalents) which are sufficiently small (see Figure 2).

11. **Location**

Location defines a geographically or ecologically distinct area in which a single event (e.g. pollution) will soon affect all individuals of the taxon present. A location usually, but not always, contains all or part of a subpopulation of the taxon, and is typically a small proportion of the taxon's total distribution.

12. **Quantitative analysis**
A quantitative analysis is defined here as the technique of population viability analysis (PVA), or any other quantitative form of analysis, which estimates the extinction probability of a taxon or population based on the known life history and specified management or non-management options. In presenting the results of quantitative analyses the structural equations and the data should be explicit.

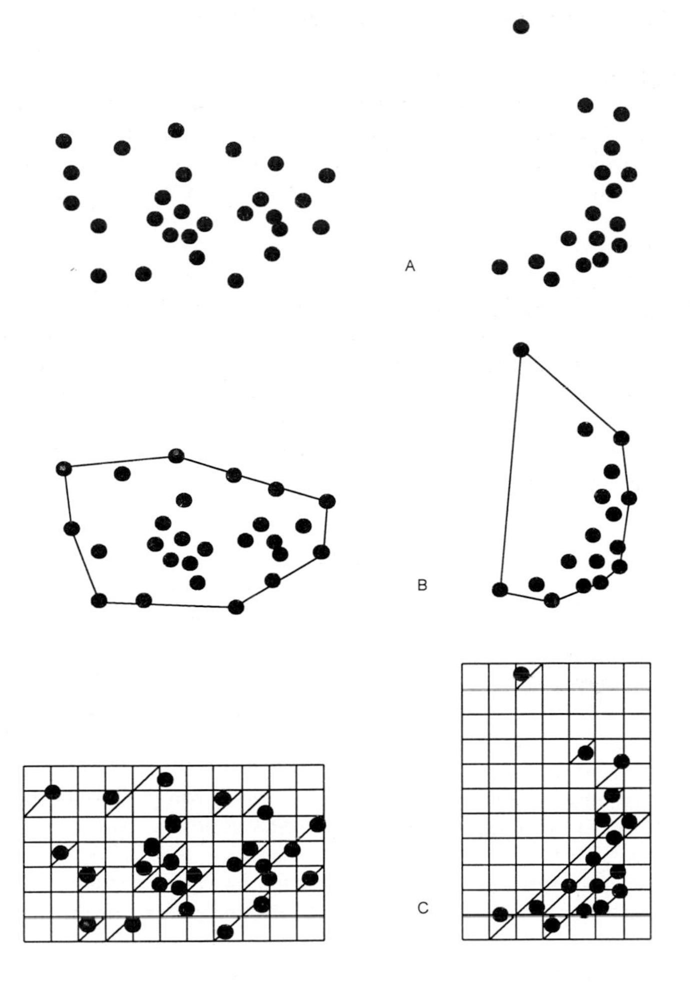

Figure 2:

Two examples of the distinction between extent of occurrence and area of occupancy. (a) is the spatial distribution of known, inferred or projected sites of occurrence. (b) shows one possible boundary to the extent of occurrence, which is the measured area within this boundary. (c) shows one measure of area of occupancy which can be measured by the sum of the occupied grid squares.

IV) The categories [1]

EXTINCT (EX)
A taxon is Extinct when there is no reasonable doubt that the last individual has died.

EXTINCT IN THE WILD (EW)
A taxon is Extinct in the wild when it is known only to survive in cultivation, in captivity or as a naturalised population (or populations) well outside the past range. A taxon is presumed extinct in the wild when exhaustive surveys in known and/or expected habitat, at appropriate times (diurnal, seasonal, annual), throughout its historic range have failed to record an individual. Surveys should be over a time frame appropriate to the taxon's life cycle and life form.

CRITICALLY ENDANGERED (CR)
A taxon is Critically Endangered when it is facing an extremely high risk of extinction in the wild in the immediate future, as defined by any of the criteria (A to E) on pages 129 and 130.

ENDANGERED (EN)
A taxon is Endangered when it is not Critically Endangered but is facing a very high risk of extinction in the wild in the near future, as defined by any of the criteria (A to E) on pages 130 and 131.

VULNERABLE (VU)
A taxon is Vulnerable when it is not Critically Endangered or Endangered but is facing a high risk of extinction in the wild in the medium-term future, as defined by any of the criteria (A to D) on pages 131 and 133.

LOWER RISK (LR)
A taxon is Lower Risk when it has been evaluated, does not satisfy the criteria for any of the categories Critically Endangered, Endangered or Vulnerable. Taxa included in the Lower Risk category can be separated into three subcategories:

1. **Conservation Dependent (cd)**. Taxa which are the focus of a continuing taxon-specific or habitat-specific conservation programme targeted towards the taxon in question, the cessation of which would result in the taxon qualifying for one of the threatened categories above within a period of five years.

2. **Near Threatened (nt)**. Taxa which do not qualify for Conservation Dependent, but which are close to qualifying for Vulnerable.

3. **Least Concern (lc)**. Taxa which do not qualify for Conservation Dependent or Near Threatened.

---

[1] Note: As in previous IUCN categories, the abbreviation of each category (in parenthesis) follows the English denominations when translated into other languages.

DATA DEFICIENT (DD)
A taxon is Data Deficient when there is inadequate information to make a direct, or indirect, assessment of its risk of extinction based on its distribution and/or population status. A taxon in this category may be well studied, and its biology well known, but appropriate data on abundance and/or distribution is lacking. Data Deficient is therefore not a category of threat or Lower Risk. Listing of taxa in this category indicates that more information is required and acknowledges the possibility that future research will show that threatened classification is appropriate. It is important to make positive use of whatever data are available. In many cases great care should be exercised in choosing between DD and threatened status. If the range of a taxon is suspected to be relatively circumscribed, if a considerable period of time has elapsed since the last record of the taxon, threatened status may well be justified.

NOT EVALUATED (NE)
A taxon is Not Evaluated when it is has not yet been assessed against the criteria.

V) The Criteria for Critically Endangered, Endangered and Vulnerable

CRITICALLY ENDANGERED (CR)
A taxon is Critically Endangered when it is facing an extremely high risk of extinction in the wild in the immediate future, as defined by any of the following criteria (A to E):

A) Population reduction in the form of either of the following:

1) An observed, estimated, inferred or suspected reduction of at least 80% over the last 10 years or three generations, whichever is the longer, based on (and specifying) any of the following:

a) direct observation
b) an index of abundance appropriate for the taxon
c) a decline in area of occupancy, extent of occurrence and/or quality of habitat
d) actual or potential levels of exploitation
e) the effects of introduced taxa, hybridisation, pathogens, pollutants, competitors or parasites.

2) A reduction of at least 80%, projected or suspected to be met within the next ten years or three generations, whichever is the longer, based on (and specifying) any of (b), (c), (d) or (e) above.

B) Extent of occurrence estimated to be less than 100 km^2 or area of occupancy estimated to be less than 10 km^2, and estimates indicating any two of the following:

1) Severely fragmented or known to exist at only a single location.

2) Continuing decline, observed, inferred or projected, in any of the following:

a) extent of occurrence
b) area of occupancy

c) area, extent and/or quality of habitat
d) number of locations or subpopulations
e) number of mature individuals.

3) Extreme fluctuations in any of the following:

a) extent of occurrence
b) area of occupancy
c) number of locations or subpopulations
d) number of mature individuals.

C) Population estimated to number less than 250 mature individuals and either:

1) An estimated continuing decline of at least 25% within 3 years or one generation, whichever is longer or

2) A continuing decline, observed, projected, or inferred, in numbers of mature individuals and population structure in the form of either:

a) severely fragmented (i.e. no subpopulation estimated to contain more than 50 mature individuals)
b) all individuals are in a single subpopulation.

D) Population estimated to number less than 50 mature individuals.

E) Quantitative analysis showing the probability of extinction in the wild is at least 50% within 10 years or 3 generations, whichever is the longer.

ENDANGERED (EN)
A taxon is Endangered when it is not Critically Endangered but is facing a very high risk of extinction in the wild in the near future, as defined by any of the following criteria (A to E):

A) Population reduction in the form of either of the following:

1) An observed, estimated, inferred or suspected reduction of at least 50% over the last 10 years or three generations, whichever is the longer, based on (and specifying) any of the following:

a) direct observation
b) an index of abundance appropriate for the taxon
c) a decline in area of occupancy, extent of occurrence and/or quality of habitat
d) actual or potential levels of exploitation
e) the effects of introduced taxa, hybridisation, pathogens, pollutants, competitors or parasites.

2) A reduction of at least 50%, projected or suspected to be met within the next ten years or three generations, whichever is the longer, based on (and specifying) any of (b), (c), (d), or (e) above.

B) Extent of occurrence estimated to be less than 5000 km^2 or area of occupancy estimated to be less than 500 km^2, and estimates indicating any two of the following:

1) Severely fragmented or known to exist at no more than five locations.

2) Continuing decline, inferred, observed or projected, in any of the following:

a) extent of occurrence
b) area of occupancy
c) area, extent and/or quality of habitat
d) number of locations or subpopulations
e) number of mature individuals.

3) Extreme fluctuations in any of thc following:

a) extent of occurrence
b) area of occupancy
c) number of locations or subpopulations
d) number of mature individuals.

C) Population estimated to number less than 2500 mature individuals and either:

1) An estimated continuing decline of at least 20% within 5 years or 2 generations, whichever is longer, or

2) A continuing decline, observed, projected, or inferred, in numbers of mature individuals and population structure in the form of either:

a) severely fragmented (i.e. no subpopulation estimated to contain more than 250 mature individuals)
b) all individuals are in a single subpopulation.

D) Population estimated to number less than 250 mature individuals.

E) Quantitative analysis showing the probability of extinction in the wild is at least 20% within 20 years or 5 generations, whichever is the longer.

VULNERABLE (VU)
A taxon is Vulnerable when it is not Critically Endangered or Endangered but is facing a high risk of extinction in the wild in the medium-term future, as defined by any of the following criteria (A to E):

A) Population reduction in the form of either of the following:

1) An observed, estimated, inferred or suspected reduction of at least 20% over the last 10 years or three generations, whichever is the longer,, based on (and specifying) any of the following:

a) direct observation
b) an index of abundance appropriate for the taxon
c) a decline in area of occupancy, extent of occurrence and/or quality of habitat
d) actual or potential levels of exploitation
e) the effects of introduced taxa, hybridisation, pathogens, pollutants, competitors or parasites.

2) A reduction of at least 20%, projected or suspected to be met within the next ten years or three generations, whichever is the longer, based on (and specifying) any of (b), (c), (d) or (e) above.

B) Extent of occurrence estimated to be less than 20,000 km^2 or area of occupancy estimated to be less than 2000 km^2, and estimates indicating any two of the following:

1) Severely fragmented or known to exist at no more than ten locations.

2) Continuing decline, inferred, observed or projected, in any of the following:

a) extent of occurrence
b) area of occupancy
c) area, extent and/or quality of habitat
d) number of locations or subpopulations
e) number of mature individuals.

3) Extreme fluctuations in any of the following:

a) extent of occurrence
b) area of occupancy
c) number of locations or subpopulations
d) number of mature individuals.

C) Population estimated to number less than 10,000 mature individuals and either:

1) An estimated continuing decline of at least 10% within 10 years or 3 generations, whichever is longer, or

2) A continuing decline, observed, projected, or inferred, in numbers of mature individuals and population structure in the form of either:

a) severely fragmented (i.e. no subpopulation estimated to contain more than 1000 mature individuals)
b) all individuals are in a single subpopulation.

D) Population very small or restricted in the form of either of the following:

1) Population estimated to number less than 1000 mature individuals.

2) Population is characterised by an acute restriction in its area of occupancy (typically less than 100 km^2) or in the number of locations (typically less than 5). Such a taxon would thus be prone to the effects of human activities (or stochastic events whose impact is increased by human activities) within a very short period of time in an unforeseeable future, and is thus capable of becoming Critically Endangered or even Extinct in a very short period.

E) Quantitative analysis showing the probability of extinction in the wild is at least 10% within 100 years.

NOTES

NOTES

NOTES